FIRST LIGHT

Also available in the Bloomsbury Sigma series:

Sex on Earth by Jules Howard
Spirals in Time by Helen Scales
A Is for Arsenic by Kathryn Harkup
Herding Hemingway's Cats by Kat Arney
Death on Earth by Jules Howard
The Tyrannosaur Chronicles by David Hone
Soccermatics by David Sumpter
Big Data by Timandra Harkness
Goldilocks and the Water Bears by Louisa Preston
Science and the City by Laurie Winkless
Bring Back the King by Helen Pilcher
Built on Bones by Brenna Hassett
The Planet Factory by Elizabeth Tasker
Wonders Beyond Numbers by Johnny Ball
I, Mammal by Liam Drew
Making the Monster by Kathryn Harkup
Catching Stardust by Natalie Starkey
Seeds of Science by Mark Lynas
Eye of the Shoal by Helen Scales
Nodding Off by Alice Gregory
The Science of Sin by Jack Lewis
The Edge of Memory by Patrick Nunn
Turned On by Kate Devlin
Borrowed Time by Sue Armstrong
We Need to Talk About Love by Laura Mucha
The Vinyl Frontier by Jonathan Scott
Clearing the Air by Tim Smedley
Superheavy by Kit Chapman
Genuine Fakes by Lydia Pyne
Grilled by Leah Garcés
The Contact Paradox by Keith Cooper
Life Changing by Helen Pilcher
Friendship by Lydia Denworth
Death by Shakespeare by Kathryn Harkup
Sway by Pragya Agarwal
Bad News by Rob Brotherton
Unfit for Purpose by Adam Hart
Mirror Thinking by Fiona Murden
Kindred by Rebecca Wragg Sykes

FIRST LIGHT

Switching on Stars at the
Dawn of Time

Emma Chapman

BLOOMSBURY SIGMA
LONDON · OXFORD · NEW YORK · NEW DELHI · SYDNEY

BLOOMSBURY SIGMA
Bloomsbury Publishing Plc
50 Bedford Square, London, WC1B 3DP, UK

BLOOMSBURY, BLOOMSBURY SIGMA and the Bloomsbury Sigma logo
are trademarks of Bloomsbury Publishing Plc

First published in the United Kingdom in 2020

Photo credits (t = top, b = bottom, l = left, r = right, c = centre)

Colour section: P. 1: © NASA/Aubrey Gemignani (t); © Howard Butler (c);
© Capella Observatory (optical), with radio data from Ilana Feain, Tim Cornwell, and
Ron Ekers (CSIRO/ATNF), R. Morganti (ASTRON), and N. Junkes (MPIfR) (b).
P. 2: © ESO (t); © NASA/JPL–Caltech (b). P. 3: © Emma Chapman (tr); © Mattis
Magg/2020; © N.A.Sharp, NOAO/NSO/Kitt Peak FTS/AURA/NSF (cr);
© Emma Chapman (b). P. 4 © Clark et al. 2011. *Science* (t); Artwork © Katie Paterson,
photo © Manu Palomeque (c). P. 5: © ESA and the Planck Collaboration (c);
© Emma Chapman (b). P. 6: © Dylan Nelson/Illustris Collaboration (t); ©NASA,
ESA, and S. Beckwith (STScI) and the HUDF Team (b). P. 7: © EHT
Collaboration (t); ©NASA, H. Fort (JHU), G. Illingworth (USCS/LO),
M. Clampin (STScI), G. Hartig (STScI), the ACS Science Team, and ESA (tc);
© NASA/European Space Agency (bc); © Northrop Grumman (b). P. 8:
© Michael Goh and ICRAR/Curtin. Inset image of a single antenna © ICRAR (t);
© Emma Chapman (c); ©Richard Porcas/1988 (b).

A catalogue record for this book is available from the British Library

Library of Congress Cataloguing-in-Publication data has been applied for

ISBN: HB: 978-1-4729-6292-8; TPB: 978-1-4729-6293-5;
eBook: 978-1-4729-6290-4

2 4 6 8 10 9 7 5 3 1

Typeset by Deanta Global Publishing Services, Chennai, India
Printed and bound in Great Britain by CPI Group (UK) Ltd, Croydon CR0 4YY

Bloomsbury Sigma, Book Fifty-eight

To find out more about our authors and books visit www.bloomsbury.com
and sign up for our newsletters

To Lyra, Cassie and Olive.
Never stop asking questions

Contents

Introduction   9

Chapter 1: Over the Rainbow   21

Chapter 2: Where is Population III?   47

Chapter 3: The Small Bang   65

Chapter 4: A Lucky Cloud of Gas   87

Chapter 5: The Dark Ages   109

Chapter 6: Fragmenting Stars   139

Chapter 7: Stellar Archaeology   159

Chapter 8: Galactic Cannibalism   181

Chapter 9: The Cosmic Dusk   211

Chapter 10: The Epoch of Reionisation   229

Chapter 11: Unknown Unknowns   261

References   277
Acknowledgements   293
Index   295

Introduction

Teach me your mood, O patient stars!
Who climb each night the ancient sky,
Leaving on space no shade, no scars,
No trace of age, no fear to die.

Ralph Waldo Emerson

In an age where we have particle colliders and space telescopes, it's hard to imagine a time when we could solve the biggest questions of the cosmos by just looking up. Look up at the night sky, and what do you see? You might be lucky to live in an area of low light pollution, so perhaps you can see the Milky Way splashed across the sky. Or perhaps there is a full Moon. Overall, though, the key characteristic of the night sky is that it is dark. Why? The littlest of words with the biggest of consequences.

For centuries, natural philosophers, physicists, astronomers and even poets wondered why the sky is dark. Their belief was that the Universe was infinitely old and infinitely large, as they had no evidence to the contrary. Olbers' paradox (named after the German astronomer Heinrich Wilhelm Olbers) states that if the Universe is infinitely old and unmoving then every direction you look in should land on a star. The problem captured the imagination of many and even Edgar Allan Poe weighed in, in his 1848 prose poem 'Eureka': 'Were the succession of stars endless, then the background of the sky would present us a uniform luminosity, like that displayed by the Galaxy – since there could be absolutely no point, in all that background, at which would not exist a star.'

The sky should be as bright as the Sun, everywhere. Well, that's not right, because if that were the case we'd have no need for street lighting. And so we wondered, for generations, what made the sky dark. To break a paradox, you break the assumptions, and in this case both assumptions of infinite age and lack of movement are wrong. Our Universe started with a Big Bang, beginning the expansion of space-time. In one fell swoop, the Big Bang requires the Universe to begin (i.e. not be of infinitely old age), and expand (i.e. move). We can solve the paradox because, even though it has been 14 billion years since the Big Bang, insufficient stars have been born for every line of sight to land on a star. Even in the deepest images from the Hubble Space Telescope we see that the galaxies account for only a small fraction of the picture, and each galaxy hosts billions of stars. Breaking the assumption of an infinitely old Universe has profound consequences. The Universe began. There were not just stars, but *first* stars, and second and third stars for that matter. What we experience now is only one stage of a much bigger cosmological lifetime (Figure 1). It's a lifetime that we can be pretty smug about understanding. We have observations of young and old stars, galaxies that are ancient and galaxies that are newly formed. We live in a time with unprecedented access to the Universe and its history, and our ability to fill in those gaps in knowledge has increased at a lightning-fast pace.

Astronomy has broadened from a necessity in ancient civilisations, through a curious hobby for the rich, to an existential science broadcast to the masses. It has become so much a daily part of our lives that, while once we celebrated the discovery of new stars and new galaxies, now we barely lift an eyebrow if we discover a new planet. We understand our Universe's lifetime, even right back to the Universe's birth story: the Big Bang. We have lots of data … but it isn't enough. Despite the exponential increase in technology and progress, there is a period in our Universe that, until recently, we had no observations of at all. From 380,000 years after the Big Bang to about 1 billion years after it, the Universe has remained in the Dark Ages. The first stars were born less than 200 million years after the Big Bang. Formerly, the Universe had been dark and empty until a simple star

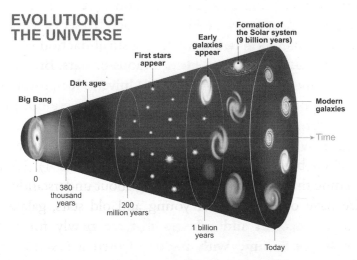

Figure 1 We are missing data from about 380,000 years after the Big Bang to 1 billion years after the Big Bang.

roared to life. A chain of internal nuclear reactions sparked into operation, producing light and heat into the barren surrounding space. Another star lit up, then another, until they speckled the sky with the dawn of first light. It was too early for planets, so there was no one around to witness that initial burst of star formation or mourn their almost immediate deaths. As the second population of stars established itself, that first generation was quickly forgotten, yet it was their intervention above all that prepared the Universe for the immense variety of structure and life that we see today. The absence of observations from the era of the first stars is alarming your local astrophysicist for two reasons.

1. Incomplete data = incorrect conclusions

First, a lack of data on this scale really matters. Missing data in any situation is a cause for concern, as it is data that we use to inform our decisions and progress our understanding. A lack of data can lead to uncertainty and misunderstandings or failings. Let us consider how aliens in our neighbourhood galaxy, Andromeda, might seek to understand the human life cycle. If our research aliens are like researchers on this planet, they are overworked and rely on the whim of funding agencies to give them the money and time to study. Our aliens don't have the resources to spend 30 years gathering data by scanning humans around the Earth, and instead will do what all scientists do when studying sizeable populations: take a sample. For example, they might take a photograph in a nursery or nursing home, or on a beach in Australia or the casinos of Las Vegas. If our aliens are really feeling the pinch of the latest research budget cuts, they might observe one location, perhaps choosing the queue of Space Mountain in Walt Disney World at random. They will take their data and walk away unaware that they have missed an extensive section of society in that sample, namely pregnant women and humans under the age of seven. Unfortunately for our aliens, with incomplete data often come incorrect conclusions. Without these sections of society, it is hard to understand the human life cycle and easy to go down the wrong path. Perhaps during a cursory literature review they read the story of babies being delivered by a stork and shouted 'Eureka!', as this fitted their data adequately. Whoops. Incomplete data? Incorrect conclusions.

In human terms, the missing cosmological data is equivalent to missing everything from the moment of conception to the first day of school, perhaps apart from a single ultrasound. It may be a small fraction of time compared to the total lifetime, but when you consider how formative these early years are for humans, it is no wonder that astrophysicists quake at this much missing data when it comes to the history of our Universe. What incorrect conclusions are we coming to about the stars around us or how the Universe is behaving now, because of this lack of data?

Your local astrophysicist is alarmed at this lack of data for a second reason too.

2. The era of the first stars is unique

The first stars are not a first edition Harry Potter, with the same story but printed on older pages. The first star is a species unto itself, a missing link that may now be extinct altogether. But surely one star is like another

star? We have one really close by after all, and it would
save a lot of effort if we could just study that one. Look
up into the night sky and you will see only a few
thousand of the several hundred billion stars in the Milky
Way. To our Universe the Sun is an unremarkable star.
Even so, generalising all those stars from looking at the
one closest to us would be like our aliens arbitrarily
observing only one human and generalising that
therefore the entire human population was called Elvis,
measured 1.8m and enjoyed peanut butter sandwiches. It
is only in the last 250 years that we have understood that
not all stars are (in fact most stars are not) the same as
our Sun.

Stars are made mostly of hydrogen and helium but we
can divide the stars we see into three populations, based
on the amount of metals within them. In astronomy
when we discuss metals we don't exclusively mean gold,
silver, platinum and so on. When we look at the
abundances of elements in the Universe, hydrogen and
helium dominate. Because of this, and as astrophysicists
are used to rounding up astronomical distances and huge
timescales, we rounded up the periodic table too. Because
I am an astrophysicist, in this book I will refer to all
chemical elements other than hydrogen or helium as
metals. For clarity I have reproduced the astrophysicist's
periodic table overlaid on the chemist's periodic table
in Figure 2.

Back to the populations. When the Universe started
out, it was predominantly filled with hydrogen and
helium. Anything heavier than that had to be created in
the hot furnaces of stars or in the energetic explosions
that ended their lives. With each generation of stars, the

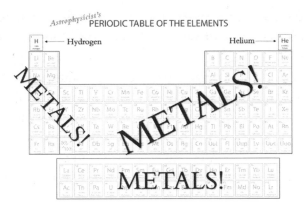

Figure 2 *The chemist's periodic table overlaid by the astrophysicist's periodic table.*

gas became more dense with metals, and the more metals would be present inside the next-generation stars that formed from that gas. The most recent generation, Population I stars, are young stars that have lots of metals inside them. They are luminous, hot and live in a galaxy's disk. Population II stars are older and have fewer metals. They reside in the centre of the galaxy or outer halo. It doesn't take much imagination to carry on down this road and ask: what about the oldest stars? The stars with no metals at all, the stars that started it all. Where are they? The first stars produced the first metals, seeding the Universe and enabling galaxies to form. They were metal-free to start with and we call them Population III stars.[*]

Population III stars were ancient beasts of mammoth proportion, up to several hundred times the mass of our

---

[*] The counter-intuitive numerical order of these populations is an artifact of their historical discovery and grouping.

Sun. They lived fast, with lifetimes of only a few million years compared to the 10-billion-year lifetimes of less massive stars such as our Sun. The same diversity of lifetimes in anthropology would be equivalent to finding an early humanoid species that aged and died only three days after birth. And yet in such short lifetimes, those stars are the ones most responsible for changing the Universe. As they roared to life, they illuminated the Universe, irradiating it and seeding it with metals that could then form stars, planets and us.

★ ★ ★

Midway through writing this book I posted a picture on Twitter that showed me working on a chapter with a frown on my face, while holding my five-week-old daughter on my shoulder. That was at around 9.00 a.m. At 4.00 p.m. I started to feel a bit funny and kept getting single-digit sums wrong while helping out with homework. At 5.00 p.m. I was struggling to breathe in the emergency department of my nearest hospital. I had developed sepsis, a life-threatening condition where an infection overwhelms the immune system, leading to multiple-organ failure, and in 20 per cent of cases in England, death. Why am I telling you this? Well, partly because I hope I've guilt-tripped the more empathetic of you browsing in the bookshop into buying the book: I nearly died writing it. Open your wallet! The actual reason I'm telling you this, however, is because one of the contributing factors to my body breaking down was my lack of exposure to starlight, or since we are on a first-name basis with our nearest star, sunlight. It turned out, after months of recovery and medical investigations, that I was deficient in vitamin D, a vitamin our bodies make with the help of UVB sunlight. We associate exposure to sunlight with the risk of sunburn, skin cancer and sunstroke, all important concerns, but a complete lack of sunlight can also damage human life. Severe vitamin D deficiency is a major contributing factor to debilitating skeletal conditions and is associated with suppression of the immune system. Now, I am a scientist so I realise that as a sample of one I cannot quantify the contribution vitamin D deficiency had in my case. Allow me to commandeer a personal dramatic story for a larger message, though. The human species depends on the Sun, not just to enable and maintain our physical habitat, but in terms of the survival of our biological habitat, our bodies, too. Our bodies are made from the metals forged in the very first stars, as well as the generations that followed. We are a machine made from the dying expulsions of stars, and we need sunlight to keep that machine running.

Interpreting the surrounding Universe is important for many reasons, and the era of the first stars has the potential to shed light not only on the early years of our Universe, but on the contemporary mysteries astrophysicists grapple with today. To understand a troubled teenager, we need to consider what happened when they were a toddler. Maybe the growing pains of galaxies 500,000 years after the Big Bang can explain why there are so many dwarf galaxies orbiting the Milky Way now.

★ ★ ★

Science is about asking questions and so I expect you have finished the Introduction with far more questions than you started with. Hopefully, by the end of the book we will have many of those questions answered... though it being science we will also probably have added more too, in the fractal way of scientific queries. We may still be stumbling in the dark, but we have established some important facts with which to start our search for first light. We know that there were first stars because the Universe began with a Big Bang, and therefore there is a first of everything. We know a lot about the history of the Universe since the Big Bang, but the era of the first stars constitutes a billion-year-long blank space in that timeline. It's not just for completeness that we want to fill in that gap – the astrophysics is exotic and worth investigation. The first stars were a species unto themselves: hot, massive and fleeting. They produced the metals (elements heavier than helium) that seed the Universe today and make up our own planet and other bodies. Learning about the baby steps of the Universe could also help us understand modern impenetrable mysteries, like how the black hole at the centre of our Galaxy got so very large. This book charts what we know about the era of the first stars, the early black holes and the first galaxies. It covers the race to discover those Population III stars, either by doing stellar archaeology or by searching for their dying signals from long ago. We will decipher clues from the Big Bang and follow the fossils of the oldest galaxies that exist in our Universe. We will need a myriad of telescopes, from the space-bound infrared to the more down-to-earth radio contraptions sitting in deserts and fields around the globe. But most of all, we will need our curiosity as we poke, prod and prise open the lid on a time that, until now, has been left in the dark. What we will see, the surprises we will unearth, well … sit back, relax and get ready to watch the Cosmic Dawn.

CHAPTER ONE

Over the Rainbow

In January 1925, US Navy navigator Alvin Peterson climbed back into the belly of an airship and realised that his cheeks, chin and fingers were frostbitten.[1] Peterson hadn't noticed this before because he had been too busy standing on top of the 200m (650ft) -long airship, steadily cranking a film camera and making the first motion picture of a solar eclipse. The *USS Los Angeles* was then the largest balloon airship ever made, built in Germany as part of the First World War reparations package. For this special voyage, scientists had commandeered it. Far below the *USS Los Angeles*, small groups of representatives of the Edison Company stood on the rooftops of apartment buildings in Manhattan.[2] Wearing long coats and fedoras, their faces stony with concentration, they struck a rather sinister picture. In each group, one person observed the eclipse and noted the degree to which the Moon covered the Sun, and whether *totality*, full coverage, was achieved. As a secondary check, another employee would look at the ground and determine whether they stood in the shadow of the Moon or not. In the streets below were amateur astronomers and schoolchildren filling out questionnaires detailing what they could see.[3] Enormous resources had gone into viewing this eclipse, from the chartering of army vehicles to the drafting of the general populace as citizen scientists. This broad engagement was engineered to improve models of the Moon's motion, and to have as many eyes as possible on the part of the Sun that was only revealed during an eclipse. As the Moon moved in front of the solar disc, one sole figure stood on top of an air balloon and was the closest human to the Sun. Alone and in darkness, Alvin Peterson was the best able to see the outermost layer of the Sun, the corona, flash into view. The corona, 'crown' in Latin, is usually hidden because the central disc of the Sun outshines it by far. Earlier eclipses had indicated that the corona contained a new chemical element called *coronium*,[4] and the 1925 eclipse provided an opportunity to study it further. However, if you check a twenty-first-century pocket periodic table, you will note that there is no coronium included. So what happened?

Eclipses and their revelations
To experience a total solar eclipse, the Moon must be in the right place in between the Earth and Sun according to the observer. The Moon's orbital plane (the surface it sweeps out as it travels around the Earth) is at an incline of around 5 degrees compared to the ecliptic, the name we give to the Earth's orbital plane around the Sun. If the two were aligned then total solar eclipses would occur once a month for any one observer on Earth. As it is, the Moon only crosses the Earth-Sun orbital plane while in between the two twice a year. A further complication comes from where in its orbit the Moon is, as its orbit is elliptical and not circular. Try sticking out your thumb and looking at it with one eye closed.

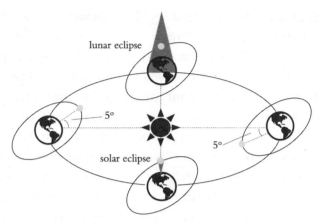

lunar eclipse

5°

5°

solar eclipse

Figure 3 *Eclipse mechanics. The ellipticity and tilt of the Moon's orbit conspire such that total solar eclipses are rare as observed from any one place on Earth.*

As you move your thumb further away from you it blocks out a smaller area of the background. If the Moon is in the elongated part of its orbit as it moves in front of the Sun, the Moon will appear too small, leaving us with a ring of fire instead of totality: an annular eclipse.

If the Moon is crossing the ecliptic, and in between the Sun and Earth, and is the right angular size to block out the Sun, then a total solar eclipse will occur and the resulting shadow will saunter across a long, thin section of the Earth as it rotates. Taking all of this into account, total solar eclipses occur somewhere on Earth around every 18 months. Any one location on Earth will experience a total solar eclipse about every 375 years.[5] Lunar eclipses, where the Moon passes into the Earth's shadow so that the Earth is in between the Sun and Moon, occur for a much greater proportion of people each time. This is because the Earth's shadow on the Moon is broader than the Moon's shadow on Earth, so when a lunar eclipse happens, everyone on the night-time side of Earth can see it.

In the past, eclipses were greeted with apprehension, their rarity lending itself to an unnatural, or mystical, explanation. They may be rare but the dance of the Earth and Moon has long been well known to some, allowing the prediction of eclipses for centuries. In 1504, when moored in Jamaica, the Italian explorer Christopher Columbus was facing a mutiny on his ship after the local population had stopped its coerced gifts of food and supplies.[6] Columbus told local leaders that God was angry at them for their reluctance and predicted (using his astronomical tables) that the Moon would soon rise with wrath. When a lunar eclipse occurred as predicted, the donations resumed. One of the most fascinating and committed attempts to predict eclipses was the Antikythera mechanism built more than 2,000 years ago in Ancient Greece.[7] The Antikythera device comprised more than 30 gears and is believed to have been the first analogue computer. The gears worked together to model movements of the Moon and Earth and predict when the two would align to create an eclipse. The machine wasn't perfect, as a sophisticated understanding of gravity is needed for the high accuracy predictions we can achieve today, but for its time the attempt was jaw-dropping. Imagine painstakingly building the device and having to wait for an eclipse to check your answer, before you added a gear to adjust it, then waited for the next eclipse. Building an understanding of eclipses is a waiting game. Over the years astronomers added ever more complexity into the mathematical models tracing the motion of the Moon, Earth and Sun and other planets in the Solar System. Even the tiny tug of distant Jupiter could influence the timing of a solar eclipse. English astronomer Edmond Halley carried out one of the first documented efforts to predict a total solar eclipse in 1715.[8] Halley is now more famous for using Newton's laws of gravity to calculate the periodicity of the comet that bears his name. However, in 1715 he used the same laws to predict an eclipse to within just four minutes. By the late nineteenth and early twentieth centuries, scientists had managed to decrease the error in eclipse estimates to a mere half a minute.

In 1925, when the Edison Company employees descended from the rooftops of Manhattan, they reported their results to Yale professor Ernest Brown.[9] He found that those standing north of 96th street saw totality, while those below did not, pinpointing one edge of the Moon's shadow to within the width of one street. Brown had also organised to measure the speed of the Moon's shadow and had speedy telegraph operators along the shadow path signal Bell Telephone Labs in Manhattan as soon as they determined that totality had begun. As signals came in from stations across the state, Brown used the timings to confirm that the shadow was moving at an astounding 5,633km/h (3,500mph). Ernest Brown was testing his new model for eclipse prediction. The Hill–Brown theory was the most complex calculation to date and predicted the 1925 eclipse to within four seconds. Oddly, *The New York Times* still placed a negative spin on that success in its headline 'Eclipse 4 seconds late here, but a brilliant show'.[10]

The complexities of the Solar System are now understood to such a degree that computers can predict eclipses to within one second, an accuracy level that is

unlikely to improve. This stagnation is not due to the
need for better computers, or other complex theories,
or more sinister-looking men on rooftops. The rate at
which the Earth spins, about 23 hours, 56 minutes and
4 seconds per revolution, is slowing down because of
the braking effect of the gravitational pull of the Moon.
Thus a day will be longer in the future, and we cannot
predict by how much because the rate is decreasing
erratically. Given the slow pace of change, our predictions
will be correct to within a second for 1,000 years or so.

Total solar eclipses are rare for any one human to see, so
it is not surprising that we continue to greet them with
wonder, and eclipse chasers fly around the world to view
them. This keen interest is mirrored within the scientific
community as well, though it's not so much because of
what the Moon hides, but because of what it reveals.

Ninety-two years after Alvin Peterson stood freezing on
top of an airship, much warmer astrophysicists sat and
watched the footage from the two National Aeronautics
and Space Administration (NASA) jets that were chasing
the shadow of the Moon. Using planes to chase eclipses is
not new. Twenty-five aircraft launched alongside the *USS
Los Angeles* in 1925, but due to the mechanical vibrations
within they obtained no useful footage. Instead, one of the
most illuminating records of the corona during the 1925
eclipse comes from an artist, Howard Butler. He positioned
himself on top of a roof in Connecticut, sketched the
prominences (gaseous features extending outwards from
the Sun) and recorded the colours to create as accurate a
painting as possible later. Our aerial and photographic
technology has since improved, much to the joy of
astrophysicists averse to inclement weather. 'The Great
American Eclipse' on 21 August 2017 was the first total

solar eclipse seen across the width of the mainland United States in almost 100 years. The NASA jets flew up to altitudes of 15,240m (50,000ft),[11] about 50 per cent higher than most commercial airliners, to avoid water vapour, which distorts our images of space in the same way that mist or fog can obscure photographs on Earth. Both the Moon's velocity around the Earth and the Earth's velocity around the Sun vary in their orbits, so the speed of the shadow and the length of time we experience totality varies for each eclipse. For the 2017 American Eclipse, a static observer on Earth would have experienced totality for only about two and a half minutes.* Interested in getting as much data as possible, the two jets chased the shadow of the Moon as it travelled through Missouri, Tennessee and Illinois. This enabled scientists to study the corona for close to eight minutes. In 2017 they weren't looking for coronium, though, instead they were trying to understand something counterintuitive about our Sun. The elusive, faint corona is hotter than the surface of the Sun itself – a lot hotter. The centre of the star where nuclear fusion occurs is around 27 million °F, or 15 million °C, and these temperatures drop off further away from the centre so that the visible surface, or 'photosphere', of the Sun is around 9,900 °F, or 5,500 °C. Then, 2,100km (1,300 miles) above the surface of the Sun, the corona boosts the mercury back up to two million °F or more. Why? How? This is odd, akin

---

* This is actually the same duration of totality as the 1925 eclipse. The longest eclipse in the twenty-first century was off the coast of Southeast Asia on 22 July 2009, at six and a half minutes long. Even the shortest of eclipses has the power to stun, though, as Virginia Woolf exhibited in her diary of the 24–second 29 June 1927 eclipse in the north of England: '… suddenly the light went out. We had fallen. It was extinct. There was no colour….We had seen the world dead'[12].

to setting your oven in New York to 400 °F, or 200 °C, and realising you have incinerated someone's cat in Alabama. The Coronal Heating Problem[13, 14] is unresolved, and this eclipse allowed scientists a rare opportunity to study the corona in high resolution for an extended period. They hoped to see what it is that causes this heating, just as their predecessors had watched Alvin Peterson's footage and wondered what coronium was. Past or present, solar eclipses offer an opportunity for us to bypass one of the largest obstacles to solar science: the Sun itself.

The wavelengths of light

What would you draw if I asked you to sketch the Sun? I would place a bet that you would draw or imagine something like a circle surrounded by radiating lines (incidentally called 'solrads' in cartoon jargon), the same as you would have done when you were a child. The Sun is one of the first things we learn to draw but our ability to depict it doesn't much improve, and understandably so since we cannot look directly at it without risking our vision. In drawing the lines extending from the central disc we show the Sun's presence, action and importance to us. Even the oldest civilisation or the youngest child understands that the Sun gives us light, warmth and, ultimately, life.

Light is the principal means by which we perceive the world but it's hard to put your finger on what it is. We can think of it as radiation that travels as a wave, with the distance between peaks in that wave called the wavelength. Light forms a continuum of different wavelengths called the *electromagnetic spectrum*.

We are used to using the different wavelengths of light in a commonplace manner, though we may not always

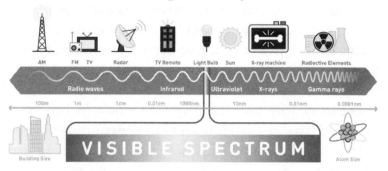

Figure 4 *Electromagnetic spectrum.*

recognise them as light. Very long wavelength light, or *radio waves*, are useful for transmitting information over long distances. For example, radio waves allow us to listen to the national news on our radio, perhaps as we use *microwaves* to cook a quick lunch. If we learn news of an escaped tiger and turn on our television, we might view the footage from police helicopters as they use *infrared* light to track the cat as it stalks into our garden. If we take a hard fall in our rush to close the patio doors, we would expect that any suspected fracture be diagnosed with an *X-ray* … and not just by a quick look under a lamp utilising *visible light*. The small part of the spectrum that we refer to as 'light' in everyday life is just the visible part that our eyes have adapted to, and we use the word *colour* to talk about the specific wavelengths in that small section.

The first person to understand that light was composed of different colours was English physicist Isaac Newton. In 1666, he aligned a glass prism with a beam of sunlight and observed a rainbow emanating from the other side of the prism.[15] I'm running low on diagram space so just look up Pink Floyd's *The Dark Side of the Moon* album cover. It had been noted before how a broken piece of glass, or a prism, resulted in different colours when held up to sunlight. What was a surprise was how, when Newton focused the light on a far-off wall, the rainbow was an oblong with the various colours spread out. Until this point it was assumed that this rainbow effect was being caused by prisms creating the colour somehow. Newton took a single colour of the split light and focused it into a second prism. When it emerged the same colour and not with the added colours a prism was supposed to create, he proved that prisms didn't just manufacture rainbows. Newton established that light can be split into different colours.

It isn't just the visible light that is split in an experiment like the prism. The German-British astronomer William Herschel,* had set up a prism apparatus similar to Newton's to take the temperature of the colours of light, but he got his highest reading from the space above the red light, over the rainbow. He called these invisible light rays 'calorific rays',[17] but we know them better now as infrared light.

I have described light as a wave so far, but this is only a half-truth. Sometimes light is better understood as a particle. This seems strange, but it is a phenomenon known as wave-particle duality, a consequence of

---

* Incidentally another astronomer who didn't mind the cold. His sister Caroline Herschel, who he originally drafted in as an unwilling assistant, wrote of her skirts freezing to the ground,[16] as she did her work. She can't have minded the cold too much in the end as she eventually found a passion in what she did, becoming the first woman to be paid as a scientist and winning the highly prestigious Royal Astronomical Society Gold Medal in 1828, at a time when women weren't even allowed to attend their meetings.

quantum mechanics. We call the particles constituting light *photons*. German physicist Albert Einstein showed that the wavelength of light was equivalent to a photon that carried a certain amount of energy.[18] The shorter the wavelength, the higher the energy the photon had. I think Newton would have been thrilled with this, as he insisted throughout his scientific career that light was 'corpuscular', or made of particles. Newton's arguments were largely incorrect, but I'm sure he'd have declared victory all the same.

In evolutionary terms, it was advantageous for us to adapt to seeing visible light as it allowed us to assess environments for sustenance or threats of danger. While infrared vision would have been useful for tracking tigers in the undergrowth, in the time we've had to evolve that genetic mutation/upgrade has not yet emerged. There are examples in other animals of more unusual visual adaptations, more akin to Superman's X-ray vision. For example, some snakes have 'pit organs' in their faces that contain a membrane able to detect infrared radiation.[19] This allows them to see visible light with their eyes, and also to get an infrared image of their prey at night. Another example occurs in reindeer, which have eyes adapted to process ultraviolet (UV) light.[20] In the Arctic, visible light is low, but the snow reflects a lot of UV light, allowing reindeer to track predator/prey urine and highlight lichen, their major food source.*

---

* So if you think about it, with the help of Rudolph, Santa can identify counterfeit money and even trace amounts of cocaine and amphetamines. An uncomfortable new dimension to Santa finding out whether you have been naughty or nice.

In astrophysics, we use the different wavelengths of light to highlight contrasting elements of our Universe. While visible light is excellent for producing stunning pictures of galaxies, X-ray and radio light are well suited for revealing hidden black holes. Take the image of the galaxy Centaurus A, shown in the colour centre-fold. On the left is the galaxy observed in visible light. On the right is a composite image of the same optical galaxy but with the radio observations added. By observing with a different wavelength of light, we have revealed a galaxy that is much larger than we had realised. Even more excitingly, we have discovered the signature of a black hole at the centre by observing those gigantic lobes of ejected material. The radio lobes are caused by a small fraction of the infalling gas around the black hole being expelled as large jets of particles. Had we just looked at Centaurus A in visible wavelengths, the black hole would not have been so obvious, whereas in radio it might as well have an arrow-shaped sign saying 'Black Hole Here'.

Taking the temperature of the stars

Let's go back and colour in our drawing of the Sun. We often depict the Sun as yellow but, as we know from Isaac Newton, William Herschel and Pink Floyd, sunlight comprises a whole range of different wavelengths. The range of wavelengths of light a star emits makes up a barcode unique to that star, a stellar spectrum, with similar types of star having similar spectra. The solar spectrum demonstrates that the Sun radiates across a range of wavelengths, with only 40 per cent being visible to the human eye. The solar spectrum peaks in the blue-green wavelengths of the electromagnetic spectrum, so yellow light is not the most common light emitted by the Sun. We don't see a green Sun in the sky for two reasons. First, that blue-green peak is not the only wavelength emitted and the light we receive is a mixture of those different wavelengths, moving closer to white light. Second, the Earth's atmosphere *scatters* light like coins in a penny-pusher arcade game, redirecting light in random directions. The shorter the wavelength (the bluer the light), the more light scatters. For sunlight, this means that the atmosphere scatters the bluer light in all directions, causing us to see it from all directions and making the sky appear blue. With the bluer part of the spectrum scattered, the red and yellow parts of the sunlight remain, making the Sun appear more yellow. At sunset, when the Sun is lower, and the light has to go through more atmosphere and even the yellow light noticeably scatters, the Sun appears orangey-red. The atmospheric scattering of sunlight likewise explains why lunar eclipses are sometimes called 'blood moons'. As the sunlight passes through the atmosphere of the Earth, it scatters practically all the shorter wavelengths, leaving only the longest, red wavelengths to go forth and reflect off the Moon and into our eyes.

We can use the overall shape of the blackbody spectrum to take the temperature of a star. The shapes of stellar spectra are very well approximated by curves called blackbody curves. A 'blackbody' is an idealised object that absorbs all incident radiation and emits in a continuous spectrum or 'continuum', a bit like someone who listens to different views, then spouts endless platitudes to please everyone. Because the Sun has no hard surface to scatter incident radiation, it does indeed

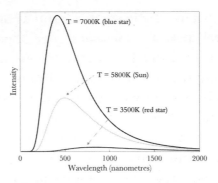

Figure 5 *The solar continuum. By comparing the shape of stellar spectra to different blackbody temperature curves we can obtain an estimate of the temperature of the surface of a star.*

absorb the vast majority of incident radiation, and as we have seen it emits light over the electromagnetic spectrum. As we can approximate the Sun, and all stars, as blackbodies, we can use the blackbody law that says that the continuous spectrum emitted depends only on the temperature. By measuring a star's spectrum we can compare it to the known blackbody curves and find which temperature curve it matches. If we do this for the Sun, we find that the photosphere of the Sun has a temperature of approximately 5,800 Kelvin (K).*

The absorption lines of coronium
The blackbody spectrum of the Sun looks like an unbroken line, continuous, but if you looked closely, you

---

* Kelvin is a measure of temperature like °F or °C. We use it because we don't like messing around with negative temperatures in astrophysics, so we mark 0K as the coldest temperature possible, called absolute zero, equivalent to -273.15 °C. To convert from Kelvin to °C you simply subtract 273.15K.

would see that it is actually made up of lots and lots of individual points. Displaying the spectrum as a continuum is great for comparing to blackbody curves and estimating the surface temperatures of stars, but the more common way in which we tend to look at spectra is as if we are looking down on that curve from above, and we achieve a barcode look. The magnificent image in the colour plates centrefold of the solar spectrum is a high-wavelength-resolution observation of the visible part of the solar spectrum. The dark lines, *absorption lines*, are due to atoms in the outer region of the Sun absorbing those wavelengths of light as they emerge from deep within the Sun. Atoms of a particular element will absorb light at very specific wavelengths relating to the energy levels that electrons are able to occupy within that atom. The gaps between these energy levels differ between elements, and therefore so do the wavelengths of light absorbed and the resulting absorption lines characteristic to an element.

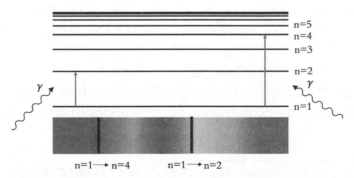

Figure 6 *Producing absorption lines. Atoms will only absorb photons (usually denoted using a γ symbol) of energies equivalent to their specific electron energy levels (top), producing distinct absorption lines in the spectrum (bottom).*

On Earth we can irradiate a specific chemical element, such as hydrogen, to see which wavelengths it will preferentially absorb and therefore what absorption lines it will produce in a spectrum. We can then look at a spectrum of a star and, by seeing which wavelengths have been absorbed, we can match with the library of experiments we have carried out on all the known chemical elements and deduce which atoms are in the Sun, if they are known at all. When English astronomer Norman Lockyer noticed an unidentified absorption line in the spectrum of the Sun in 1868,[21] he named the previously unknown element responsible *helium*, after the Greek *Helios* for god of the Sun. It would be another 27 years before helium was discovered on Earth.

During an eclipse the solar spectrum looks very different, and because the main disc of the Sun is blocked out, we just see the light from the corona. When scientists first looked at the coronal spectrum during a solar eclipse on 7 August 1869, they found a strong absorption line at a wavelength of 530.3nm, which didn't correspond to any of the experiments they had performed on Earth: no known element produced an absorption line at that wavelength. Eventually it was assumed that this must indicate a new element in the Sun's atmosphere, which they named *coronium*. The pursuit of this element would last decades and was one of the reasons why Alvin Peterson stood on top of a dirigible in 1925, freezing his fingers to further our understanding of this new element. It wasn't until the 1930s that it was realised that this absorption line was not due to a new element at all, but was instead a different form of the already familiar element, *iron*. Iron usually has 26 electrons and is associated with a well-known set of absorption lines. If we take away some of those electrons (we *ionise* the iron), then those absorption lines change. The absorption line linked to the mythical element coronium was identical to the absorption line produced by an iron atom that has had 13 electrons removed. Removing electrons is no easy job and requires very high temperatures to create high-energy collisions capable of dislodging electrons. To remove 13 electrons requires blisteringly high temperatures, indicating that the corona is millions of degrees Kelvin compared to the comparatively chilly 5,800K photosphere. So the Sun does indeed get hotter as you venture into the corona, an entirely unintuitive situation. Perhaps the high temperatures are produced by complicated magnetic fields violently throwing charged particles up from the photosphere into the corona, or perhaps bomb-like explosions called nano-flares throw heat up into the atmosphere. The theories are being narrowed down, but the Coronal Heating Problem remains an unsolved quandary.

★ ★ ★

It might disappoint you that I haven't been able to tie up the Coronal Heating Problem in a neat and satisfying way. It's annoying not knowing the answer, isn't it? And surprising considering the resources we command. We have built probes that have travelled all the way into interstellar space, and Hubble can see well outside the Milky Way to a multitude of other galaxies. We understand how our Universe evolved, even right back to having a solid theory for its creation, the Big Bang. Yet we still race to exploit every bit of scientific understanding we can from seven minutes of totality, just as scientists were doing in 1925, albeit now with more advanced technology. We add to our knowledge in one sense (the corona isn't made of an unusual element – it's just really hot!), then add to our questions in another (wait, how is it that hot?!). I will admit that the Coronal Heating Problem was fairly new to me at the time of the Great American Eclipse. As an astrophysicist studying the first stars, it hadn't ever occurred to me to pay much attention to the nearest star – the mundane, predictable Sun. I took for granted that because it was close, we had it sorted. I was fascinated and delighted to discover that there were still mysteries that caused us to utilise the ultimate serendipity of our Solar System: eclipses.

It is an amazing coincidence that during a total solar eclipse, the Moon covers up the Sun's disc like a lens cap. The Sun and Moon are the same angular size on the sky, by pure chance. The angular size of an object depends on how big something is and how far away it is. Imagine that we are looking at cows. If we notice that a toy cow, 10 times smaller than a real cow, appears the same size as a real cow, we know the real cow has to be 10 times further away. In our case, we observe the Sun and Moon as the same angular size. The Sun is about 400 times further away than the Moon but it is also about 400 times bigger than the Moon, so they appear to be of the same angular size. What makes eclipses even more precious is that they have an expiry date. I mentioned much earlier that the gravitational pull of the Moon on the Earth causes a braking action. This affects our ability to predict eclipses because it slows the Earth's rotation unpredictably over time. The other effect of this is that the Moon is moving away from the Earth by about 4cm (1½in) per year ... our toy cow is moving further away, ruining the illusion of similarity. The total solar eclipses we enjoy today will not last forever, and if any humans are around 650 million years from now, they will only know partial eclipses. Still, it's a long way off and considering that human ancestors appeared on Earth only a few million years ago, perhaps we will have evolved X-ray vision and will be viewing the Universe in a way that makes eclipses redundant.

We are now in a time when we aren't restricted to just waiting for the next eclipse. There is, for example, the coronagraph, a telescopic attachment that blocks out a lot of the photosphere. And just over a year after the Great American Eclipse, NASA launched a revolutionary new mission called the Parker Solar Probe.[22] For seven years, this probe will make its way towards the Sun, reaching to within 3.8 million miles from its surface, an unexplored area. This may not sound very close but as we have seen, the Sun's surface is not the 'edge' of the star. The corona extends far into space and it's difficult to define an end to it. The hot, energetic atmosphere results in a stream of particles called the solar wind, which encompasses the eight planets of our Solar System. This means that while the hot core of the Sun is far away, we on Earth are living well within its atmosphere. The solar wind we experience on Earth has had to travel for several days over 93 million miles to reach us on Earth, getting distorted and changed by all the other particles in the way. This makes studying the inner corona like birdwatching through a heavy fog from 100m (328ft) away. To understand it, we are going closer to get through the fog and will be standing only 4m (13ft) away. Parker will take high-resolution imagery and watch for those elusive nanoflares and magnetic field lines, helping scientists understand just how our star can reach such high temperatures so far away from the internal heat source. Interestingly, in being that much closer to the Sun, Parker will also see events such as solar flares on the Sun before those of us watching on Earth. Parker will see a solar flare 20 seconds after the actual event, while those of us on Earth will see it a full eight minutes later – that is because of the finite speed of light.

The finite speed of light

The Danish astronomer Ole Rømer came to the realisation that light did not travel at infinite speed by observing eclipses. During the 1600s, one of the pressing scientific questions was how to establish your longitude (east-west position on the Earth's surface in degrees) when sailing the high seas. We know that the Earth revolves through 360 degrees in 24 hours and so if you know the time difference between two points, you can determine the longitudinal difference too. These days we turn on our laptops and our clocks can synchronise no matter where we have travelled in the world. But how to work out the time difference in the 17th century? What was needed was an event that happened at predictable times of the day. You might be familiar with the scene early on in Disney's *Mary Poppins*, where every day at 6.00 p.m., the eccentric neighbour sets off his cannon. Anyone within a decent radius of Cherry Tree Lane would know that when they heard that sound it was 6.00

p.m., at least by the Major's watch. The Italian astronomer Galileo Galilei suggested a similar idea on a much grander scale. He needed an event that could be seen from anywhere on Earth, so he looked to the skies and found one in the Jovian system (the system encompassing Jupiter and its moons). As the Jovian moons orbited Jupiter, they disappeared from observation when they travelled behind Jupiter. When a moon had travelled far enough around, it would pop out from behind Jupiter, an event known as emergence. Because the moons orbited Jupiter regularly, Galileo realised that by observing the times of an emergence you had a celestial clock,[23] an inter-planetary cannon. This idea was tricky to implement on a moving ship at the time, but in 1668, Italian astronomer Giovanni Domenico Cassini published his tables of the Jovian moon movements, providing a timetable for people to set their clocks by. For example, sailors could look up that on 10 November they expected an emergence of Io at noon, as observed in Naples. You therefore knew that, at emergence as viewed from anywhere else in the world, it was noon in Naples. You could then determine the time difference, and therefore longitude, by comparing the time to when you measured local noon (the point the Sun is highest in the sky). At first, all the predictions seemed spot on, to at least half a minute, but as the year went on emergence events for Io were occurring about 10 minutes later than anticipated. Cassini remarked that this could only be because of either a finite speed of light, or a variation in the diameter of Jupiter. Cassini himself settled on the latter opinion, or at least opposed the other opinion, but his colleague Ole Rømer instead pursued the former.

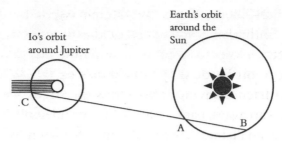

Figure 7 *The finite speed of light. Adaptation of a figure from Rømer's 1676 work[24], demonstrating that the light will take longer to traverse a longer path. Light takes longer to travel from the emergence of Io (C) to Earth at point B in its orbit, than point A.*

To understand why Rømer was correct, look at the sketch of the Jovian-Sun-Earth system in Figure 7. Behind Jupiter there is the shadow cone that Io enters and leaves as it orbits Jupiter. Observing from point A, you see the emergence of Io (C) at a certain time, say noon. The next day, the emergence still appears to be roughly on time, say 12.01 p.m. But take a measurement some time later, when the Earth has moved around in its orbit to point B, and the emergence arrives many minutes later than expected. Rømer realised that due to the increase in distance between the Jovian system and Earth, the light from Io had further to travel. In our figure, C to B was much further than C to A. If the speed of light was infinite, this wouldn't matter – light would cover any distance instantaneously. What the delayed eclipses showed was that light took a lengthier time to travel a longer distance. The speed of light is finite, but it is not slow. In everyday life we view it as instantaneous. The speed of light implied by Rømer's work on the Jovian moon eclipses implied a speed of light of 212,000,000m/s (metres per second).[24,25] This was an unthinkably fast speed at the time (and still is), but pretty close to what we can now measure in a laboratory on Earth: 299,792,458m/s exactly. When we notice a friend waving from across the street, we know that we can wave back before they wander off because there is a small, human-scale distance between us. We see our friend wave with a delay from the actual event, but the lag is only 0.00000003 seconds, so it's instantaneous to our eye. But when we scale up to the celestial scale, we find that our friend who now lives on Jupiter's moon Io can wave at us, but we will only see them do it half an hour later (by that time they will think us rather rude for not having waved back).

Due to the finite travel speed of light, the further away we observe something, the earlier we are observing it at. When we look at the surface of the Sun, the solar flare we see is eight minutes old.* When we look at one of our closest stars, Alpha Centauri, the light is more than four years old. What brings this incredible fact home for me is that if Earth is indeed being observed by alien researchers in the Andromeda system, they aren't watching us ride Space Mountain. Instead, they are writing down that this third rock from an average star harbours species such as *Paranthropus aethiopicus* and *Australopithecus africanus*,

---

* The interior of the Sun is a seething mass of agitated particles and for a photon to fight its way from the core to the surface can take hundreds of thousands of years[26]. It is worth remembering that while the light may have left the photosphere eight minutes ago, it will have left the core a touch longer: closer to 50,000,000,008 minutes ago.

archaic ancestral branches of humans that lived more than 2.5 million years ago. Andromeda is so far away that we see it as it was 2.5 million years ago, and vice versa. With greater distances involved, light becomes a gateway to looking back in time.

We are actually already familiar with this concept as it occurs in routine life. Every time you hear thunder on a hot summer's night, you are hearing an atmospheric event that occurred seconds before. When the thunder cloud is right above you, it might seem to clap at the same time that you see the lightning. As the cloud moves away, there is a greater distance for the sound and light to travel, but the light is almost a million times faster, so there is an increasing delay between seeing the lightning and hearing the thunder. We can also ask what would happen if light had a much slower speed. If light only covered 1m (about 3ft) in a few seconds, then we could look at the Moon now and see Neil Armstrong taking his first steps for humankind. It sounds fantastical, but it is this property that we are exploiting on cosmological scales. We are building a telescope to observe signals that have taken a whopping 13 billion years to arrive at Earth. This is one way in which we are searching for the first ever stars, by looking back in time to watch them form, live and die.

★ ★ ★

This chapter has been about light – and its absence. The many varieties of light form the electromagnetic spectrum, a continuum of wavelengths from radio, through visible, to gamma rays. These different wavelengths of light can be used to expose different elements of our Universe, just as X-rays are used to reveal a human's inner workings. The inner workings of the stars are revealed by the pattern of light they emit: their stellar spectrum. The shape of the intensity of that spectrum divulges its temperature. The spectral pattern of dark lines reveals what chemical elements are absorbing photons in the stellar atmosphere, creating dark absorption lines in the spectra. The speed at which light travels may seem fast, but over astronomical distances that finite speed introduces a time delay. We see the Sun eight minutes in the past. To see the first stars, then, we might just look far away enough that the light has taken 13 billion years to reach us. This requires us to look over a distance far larger than our Solar neighbourhood, farther than the extremities of our own Galaxy, even than the Local Group of galaxies, and into the depths of the Universe.

I was flicking through a book in my university library a while ago and saw a peculiar dedication. In 1964, astrophysicist George Gamow published his new book, *A Star Called The Sun*.[27] Within, he wrote, 'It is only natural that I dedicate my new book to the memory of the old one', his 'hopelessly obsolete' book, published only 24 years earlier. In the spirit of that dedication, after exploring the background of our own nearest star and understanding how and what we know, we now move forwards and prepare ourselves to encounter what we have never seen before.

CHAPTER TWO

Where is Population III?

It must have been 1997, or perhaps 1998, when I learned
how to ninja-roll under a store security shutter. Every
Saturday for several months I, and a pack of other
children, would wait outside our local department store.
At the first sound of the shutter rolling up we would all
run, jostling to roll or crawl under the shutter first. Our
first challenge over, we would jump up, turn left and run,
really run, down the store and into the toy section. Just
by the till there was a large wicker basket, full of Beanie
Babies. We would grab what we could, throwing out the
useless common varieties and releasing adrenaline-
fuelled whoops of delight when we found the rarer
bears, like the prized Britannia Beanie Baby, a simple
brown stuffed bear with the Union flag embroidered on
its chest. Madness. Not a time in my life that I am proud
of. I'm even less proud of the streak of capitalism that I
showed when selling the same bears at the outdoor
market for around five times what I had paid for them,
keeping one if I didn't have it in my collection. Luckily
for me, my parents were only gently supportive of my
obsession, allowing me to spend (or earn) my money as
I wished, but not getting involved themselves. The
Beanie Baby bubble burst a year or so later, rendering
my collection worthless. It wasn't a huge deal for me,
unlike for those who had invested thousands of dollars

or more into the soft toys. One family spent more than $100,000, aiming to use the profits to fund college tuition.[1] The manufacturer's strategy of 'retiring' certain designs at the height of their popularity drove the Beanie Baby bubble, creating scarcity and driving up demand and price. All collections have their rarer items, the most prized members of the gallery. If you find the right mint-condition stamp in your grandparents' stamp album, you might earn half a million pounds. Investment is only one motivation for collecting, though. There are those who appear to collect for pleasure alone. Everyone has an uncle with a frog collection, or a grandma with a novelty teapot on every shelf. Look hard enough and you can find a human who has a collection of pretty much anything: traffic cones, tin boxes, Barbie dolls, back scratchers. In Devon you can pay to walk around several acres of woodland filled with more than 2,000 gnomes. The minute we own something it becomes more valuable to us – we make connections with objects almost instantly and are loathe to give them up. Split a test group into two and give the groups differing objects (it doesn't matter what: money or a lottery ticket, chocolate or a mug, a mug or a different mug); they will treasure what they have been given, showing a reluctance to trade independent of what they have been given.[2] This is called the endowment effect. There's something in us that needs to classify objects, own them and keep them. British physicist Ernest Rutherford famously said that 'All science is either physics or stamp collecting', implying derision on, for example, the practice in biology of classifying the animal and plant kingdoms into species, genera, families and so on. In the same era, the early 1900s, American astronomer Edwin Frost dreamed of a system dividing the stars into phyla and classes: a kingdom of stars. We haven't quite reached the complexity of the biological kingdoms, but by applying the principles of taxonomy to the stars we have been able to draw paradigm-shifting conclusions about their constitution.

Classifying the stars

When we began to classify the world around us, we started simple: Us Down Here and That Up There. The cultural significance of astronomy took root in its role in myth and legend. Ancient Egyptians revered the Sun god Ra above all others, and depicted him as the creator of all life. On the same continent, the San people of the Kalahari Desert fear the Sun as a bringer of death. They survive searing heat by digging out hollows, urinating in the sand, then lying still beneath a layer of sand and waiting out the tortuous heat.[3] The groupings of stars were the storage sites of stories to teach, warn or just entertain. In Ancient Greece, Orion was a great hunter, pursuing the seven sisters of the Pleiades. In one Inuit legend, the Belt of Orion is a sledge, pursuing a bear represented by Betelgeuse.[4] In Australia, Orion's Belt is reduced to representing a saucepan, with the dagger as the handle. Even today, the practice of shifting responsibility for certain events or life choices based on one's star sign, or the influence of a planetary alignment, is rife. The practice of astrology split from the scientific pursuit of astronomy long ago, when it became clear that there was no evidence that the movements of planets and stars could influence our actions on an individual level.

The scientific pursuit of astronomy was born out of two human needs: time reckoning and navigation. The Sun and Moon were used to mark the passage of time by all cultures, and Polynesians used the rising and setting of the more distant stars to navigate in the Pacific Ocean long before European sailors held astrolabes and sextants. While stars were drawn, grouped and charted for thousands of years, their division into groups based on constitution, temperature and size would have to wait for the advent of *stellar spectroscopy*.

The first attempts to classify the stars scientifically began in 1863, with Italian astronomer Angelo Secchi dividing them into five groups, or Secchi classes, depending on how their spectra looked. Some stars were grouped together because their spectra had lots of absorption lines, others because they had fewer lines, or broad lines, and so on. The groups loosely aligned with colour: red stars seemed to have similar spectra, blue stars a different kind of spectrum.

The 1890s saw the industrialisation of stellar classi-fication, not with the use of computers as we know them, but with the use of women, who were referred to as 'computers'. The cheap labour women provided allowed Williamina P. Fleming to get a job as an assistant to the Harvard astrophysicist Edward C. Pickering. Fleming evaluated stellar spectra and assigned them a letter according to the strength of the hydrogen absorption lines. They began by giving the stars with the most hydrogen the letter A, those with slightly less hydrogen B and so on down the alphabet, up to Q. This was rather cumbersome, and a later assistant of Pickering, Annie Jump Cannon, reorganised the system in 1901 in order of temperature, merging

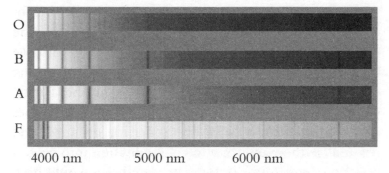

4000 nm   5000 nm   6000 nm

Figure 8 *Stellar spectra. Example spectra for the O, B, A and F stellar classifications. The absorption lines differ in number and width.*

similar classes.[5] She was quite the computer, classifying hundreds of stars per hour until eventually the alphabetical classes were reorganised into 10 classes: OBAFGKMRNS, which astronomers traditionally remember as Oh Be A Fine Girl Kiss Me Right Now Sweetheart.[*] Each of these classes is subdivided into finer numerical divisions of temperature, so that our Sun is a G2 star.

The Harvard computers categorised hundreds of thousands of stellar spectra based on the strength of absorption lines associated with hydrogen, and we moved from simply wondering at the stars to categorising them like a stamp collection. The science behind why these groupings seemed to occur was uncertain, though. Why did stars have different absorption line patterns? Whatever the strengths of the distinct lines, in the early 1900s, scientists were confident that the existence of the same lines in each spectrum, albeit at different strength, pointed

---

[*] The author of one textbook (R. J. Tayler) helpfully reassured readers that 'if this is thought sexist, Guy can replace Girl'.

to the fact that all stars were made of the same elements. More than that, scientists noted that the presence of elements such as iron mirrored the constitution of our own planet. When scientists observed metal lines in the solar spectra, they concluded that, because the same elements were present both on Earth and in the solar spectra, the bodies must be composed of the same material. They put the different intensities of the lines down to a difference in temperature. In 1925, our view was that, if heated enough, the Earth would provide the same spectrum as the Sun. Stars were just really hot planets. That year, Cecilia Payne-Gaposchkin wrote a PhD thesis that begged to differ or, more accurately, quietly provided evidence that this view was incorrect, before stating that the evidence was so fantastical that it was probably wrong and actually it might be better to just ignore it after all. This is a rare example of a PhD thesis providing a ground-breaking shift in understanding, though it was discounted at the time.

The widening of the horizon

Cecilia Payne-Gaposchkin was, in the words of many of her contemporaries, one of the truly great astronomers of the twentieth century and, more intriguingly, in the words of her daughter, an 'inventive knitter'.[6] Born in 1900 in Wendover, Buckinghamshire (a mere 8.52km/5.3 miles from my birthplace), Payne-Gaposchkin's interest in astronomy began at an early age. On seeing a meteor in the sky at five years of age, she announced her intention to be an astronomer, remarking that she had better do it quickly lest the research ran out by the time she had grown up. Her first real love was botany, adopted from her step-great-aunt Dora, and she would spend hours walking through the Chilterns, identifying rare plants, cataloguing unusual flora and thinking deeply about the scientific basis behind the incredible diversity of plants. She attended Newnham College at Cambridge University on a full scholarship, intent on following her interest. It turned out, however, that no one else was thinking quite as deeply as she was about botany, and she quickly became disillusioned with the restrictive curriculum, and lack of enthusiasm and questioning involved in her course. It's not surprising, therefore, that her head was turned to the physics building where greats such as Rutherford were redefining the composition of the Universe, defining what an atom was and whispering about a new idea called quantum mechanics. It was a dynamic time to be a physicist, and if you couldn't think deeply about the nature of the physical Universe, you might as well get out. Payne–Gaposchkin soon began to study more physical sciences.

Transferring to such a male-dominated field in the 1920s came with its own set of challenges to overcome. She remembered having to walk into every lecture theatre to the sound of stamping feet. Depressingly, I heard Northern Irish astrophysicist Jocelyn Bell-Burnell say the same of her time at university in the 1960s. Payne–Gaposchkin persevered, however, enduring the wrath of one practical lab demonstrator who shouted at the women to take off their corsets as they would interfere with the electromagnetic experiments. It was in this context that she received a last-minute free ticket to attend a lecture by Arthur Eddington regarding his trip to Brazil to view the 1918 eclipse and verify the theory of relativity. Eddington's lecture on relativity had a profound effect on Payne-Gaposchkin.

> The result was a complete transformation of my world picture. ... When I returned to my room, I found I could write down the lecture word for word ... For three nights, I think, I did not sleep. My world had been so shaken that I experienced something very like a nervous breakdown. ... I was done with biology, dedicated to the physical sciences, forever.

Eclipses pop up again and again in this book because of their singular ability to inspire, despite obscuring the very object we aim to study. The 1918 eclipse gave us not only a confirmation of relativity but resulted in one of the greatest minds in astronomy turning her head from the plants on the ground, looking to the stars and revolutionising our understanding of the Sun.

Payne-Gaposchkin's dogged pursuit of astronomy led her through Cambridge University and, when she was told job opportunities in England were not available for women astronomers, corset or no corset, on to Harvard, in the United States. Truly a renaissance woman, Cecilia Payne-Gaposchkin was fluent or semi-fluent in several languages, including Icelandic. She was proficient in classical music, and started up her own orchestra at the Harvard observatory upon finding that there wasn't one.*

---

* This is probably in large part due to the influence of Gustav Holst, who was a teacher at a school she attended in her formative years. It's quite fitting that even in her musical passions she found herself touched by astronomy, Gustav Holst being the composer of *The Planets*.

She was dedicated and thorough in everything she did. On hearing of the persecution of a fellow scientist in Germany, she aided his escape to the US. When the US Government asked people to grow their own food for the war effort, she didn't only plant a few potatoes in the back garden – she bought an entire smallholding and grew vegetables, culled turkeys, pickled everything she could get her hands on and donated some 50,000 eggs from her chickens over the wartime period. Cecilia Payne-Gaposchkin did nothing by halves.

These achievements and eccentricities pale in comparison to her scientific achievements. Usually a PhD is a training programme, involving getting up to speed and developing an expertise in a niche. Payne-Gaposchkin dived into her PhD with characteristic dedication, working for weeks and months at a time without rest, so much so that her own mother wrote to her PhD supervisor asking that she be forced to take a holiday. Payne-Gaposchkin built upon the work carried out previously on building a stellar classification system. Her breakthrough was applying new theories on the ionisation states of the atom to stellar spectra. She realised that the strength of the absorption lines wasn't a direct indication of the abundance of an element but a sign of the ionisation state of that element. Atoms comprise a nucleus surrounded by electrons. When heated, or under pressure, electrons can be stripped away from the nucleus in a process called ionisation. The resulting atom, an ion, will produce different absorption lines from its more complete counterpart. By applying ionisation theory to the stellar spectra, she could make more accurate estimations of the relative abundance of the different chemical elements present in stars. She spent two years painstakingly identifying the distinct lines present in spectra and understanding what ionisation state they indicated that atom was in. What Payne-Gaposchkin's results appeared to show was that, while the Sun did appear to contain the elements also found on Earth, such as carbon, hydrogen and helium, the relative abundances of those elements were completely different. The stars appeared to contain about a million times more hydrogen than the mere whiff of metals that her results showed. This was so odd to the astronomers of 1925 that despite all the evidence, it was put down to a misapplication of theory to observation. A fellow astronomer, Henry Russell, was 'convinced that there is something seriously wrong with the present theory. It is clearly impossible that hydrogen should be a million times more abundant than the metals'. Payne-Gaposchkin wrote up the results with the qualifying statement that the stellar abundance for hydrogen and helium 'is improbably high, and is almost certainly not real'.[7,8] Even with the qualifying statement the publication of such unexpected results damaged her reputation, with one potential boss stating that he couldn't consider hiring her firstly because she was a woman, but also because he was 'rather disappointed' to read her paper. This is not to cast blame on Russell or admonish Payne-Gaposchkin for not standing her ground. The result was bizarre in the context of the day, and even the most level-headed scientists can be distracted from evidence when it clashes so completely with their embedded world view. One only needs to look at the best-known physicist of all, Albert Einstein, and see the lengths he went to avoid the conclusion that the Universe is expanding.

Payne-Gaposchkin moved on to other research, but over the next few years other complementary evidence piled up to support her findings. In only a few short years, Russell published a paper announcing that stars were indeed mostly made of hydrogen. Sadly, Payne-Gaposchkin received little credit for her pioneering work on the subject at the time, and there was no mention of Russell's own insistence that her results were wrong. According to her daughter, Payne-Gaposchkin never minced her words or hid her bitterness or jealousy well, and she could always be relied on to sum up how she felt about a situation:

> Young people, especially young women, often ask me for advice. Here it is, valeat quantum. Do not undertake a scientific career in quest of fame or money. There are easier and better ways to reach them. Undertake it only if nothing else will satisfy you; for nothing else is probably what you will receive. Your reward will be the widening of the horizon as you climb. And if you achieve that reward you will ask no other.[9]

Thanks to Payne-Gaposchkin and Russell, the astronomical world came to terms with the fact that stars are made mostly of the lightest element in our Universe, hydrogen.

Family photographs

When we look at how the colour and luminosity of a star differ, the stars form nice, neat groups that have led to us understanding stars not as static beings but as evolving entities. A Hertzsprung-Russell (named after the same Henry Russell; H-R) diagram represents each star with a point on a graph where the star's colour tells you where to place the star on the horizontal coordinates, and the star's luminosity tells you where to place the star on the vertical coordinates. If you choose a group of stars and plot each of the stars on the H-R diagram, you find that the stars aren't distributed equally across the plot but instead form clusters and lines. This is because when we view a star system we view lots of stars, all at different points in their lifetimes. They have life stages and their attributes lend themselves into division, into families and generations. Payne-Gaposchkin was known for referring to particular stars as her friends and anthropomorphising them somewhat – and I can understand why. Looking at a stellar system, you are in fact building up an instantaneous shot of that star system's evolution. Because we are so familiar with how a human life progresses, when we take a photograph of an extended family we realise who are the new humans, who are the old ones and who might be more closely related. We can group the humans into rough groups: babies, children, teenagers, young adults, and middle-aged and senior citizens. However, when looking at a group of stars it is hard for us to understand which star might have grown up with which other star and which star is on its last legs, but the different age groups pop out as these different clusters and tracks on an H-R diagram. Each of the tracks and groups represents a different stage of stellar life. Describing a star in any static way feels wrong to me. A star is an evolving entity, from

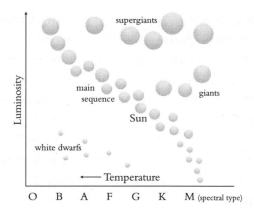

Figure 9 *The Hertzsprung-Russell diagram. In young stellar systems the majority of stars lie on the main sequence. In much older systems most of the larger mass (hotter) stars have evolved off the main sequence, turning into giants or white dwarfs.*

the first moment of collapse in a gas cloud, to its often violent demise.

The H-R diagram deconstructed at its most basic level tells us that there is a main sequence running diagonally across the middle of the plot that most stars spend 90 per cent of their lives on. Where a star lies on the main sequence depends on that star's mass or equivalently temperature, with our Sun lying about one-third up from the bottom of the main sequence. A star like our Sun will stay on the main sequence for about 9 billion years, of which we are approximately 4.6 billion years in. From the main sequence, an intermediate mass star migrates to the red giant phase, its size inflating to the extent that when our Sun is in this phase, it will just about reach Earth's orbit. What happens then depends on the mass of the star. If it's towards the lower end of the scale, its power source will run out and a small remnant known as a white dwarf will remain. If it's on the chunkier side it will explode in a supernova, which will rip itself apart and release as much energy as all the other stars in the galaxy in a sudden burst. If the core of the star survives the explosion it might settle into perpetual retirement as a neutron star, a star so dense that the mass of two Suns is packed into a sphere the size of a small city. If the surviving star remnant is particularly massive, then the remnant star collapses entirely in on itself, forming a black hole. The speeds at which stars migrate along these pathways are governed by their masses. More massive stars burn up their fuel quicker, resulting in shorter lifetimes. Thus, if we observe a very old system, then most of the high-mass stars will have disappeared off the high-mass end of the main sequence, and entered or finished the red giant phase. Just as in our family photo, if we see one from 30 years down the line, it's likely that a few of the older members of the family will have entered their senior years or, sadly, are no longer in the photo at all.

In 1944, Walter Baade noticed that he could divide stars into two types, with very different H-R diagrams, depending on their locations in the galaxy.[10,11] A spiral galaxy is usually comprised of a central concentration of stars, the 'bulge', a 'disk' with a spiral arm structure and an encompassing 'halo' of more diffuse stellar structure. Baade found that stars in the solar neighbourhood (i.e. the disk of the Galaxy) formed a different H-R diagram to the clusters of stars far off in the Milky Way halo, and these same diagrams also emerged from data drawn from the disk and halo of the Andromeda galaxy. Baade was the first to publish the idea that Type I stars were young and Type II stars were old, by observing that older

Type II halo stars will have evolved off the main sequence, creating a different H–R diagram from the younger Type I stars in the disk.[12] A main distinguishing characteristic of the stars in the two groups appeared to be the age and strength of the metal lines in the spectrum, but a physical understanding of this separation was lacking.

Cecilia Payne-Gaposchkin's work had convinced the scientific community that the Sun was different in composition from the Earth and predominantly made of hydrogen. The new status quo until the 1950s was that the components of the Sun, though different from the Earth, were then the same as every other star. Different line strengths in different spectra were erroneously understood to be purely the result of stars at different temperatures. In the 1950s, a metal-poor star was discovered with such faint metal lines that it couldn't be explained by a mere temperature difference compared with the Sun. A whole new field was born, called stellar archaeology, which we hear much more about in Chapter 7. Weaker metal lines indicated lower intrinsic metal content, and coupled with the theory that there were two different populations of young and old stars, this implied that the metal content of the Universe increased over time. Older Type II stars were more metal-poor than the young Type I stars. We understood for the first time that these heavy metals were produced inside the stars, spreading into the Universe as the stars exploded. The younger stars form from gas already seeded with metals, giving them a higher metal content. If it weren't for the stars creating carbon, iron and nitrogen, we wouldn't exist at all, or as American astrophysicist Carl Sagan put it, 'We are made of star stuff.'

Stellar populations

We still use these types today, except that we refer to them as Population I and Population II stars. Population I stars are young, metal-rich stars most commonly found in the disk of the galaxy. Our Sun is a Population I star. Population II stars are older, more metal-poor stars, more often found in clusters in the diffuse, encompassing halo of our galaxy, or in the central concentration of stars, the bulge. Over the course of the first half of the twentieth century, we saw our understanding of the stars evolve from them being entities similar to Earth, entities similar to the Sun and finally a diverse collection. There are stars of different colours, structure, size and constitution. There are young ones, old ones, common ones and rare ones. The rarest of all is the star we need to complete our collection. Over the last half of the twentieth century, astronomers realised that the Big Bang could not synthesise all the elements we saw around us, and that stars must have a role. This conclusion also meant that there must have been a more metal-poor, in fact almost zero-metal gas, right at the beginning of the Universe. Population III then, is the class of star that formed from the primordial metal-free gas. These are the first stars, the metal-free stars, and these terms are used interchangeably throughout this book. Population III stars were, and I suppose still are, a theoretical concept. They follow logically from our observations and understanding of the chemical evolution of the early Universe, but we are yet to have our theories confirmed in observation. Astronomers have been searching since their conception as an idea, but so far the stars have evaded examination.

★ ★ ★

There is a joke in academia that if a paper has a question in the title, you can save yourself reading all the waffle as the answer will be always be 'no'. In 1981 American astrophysicist Howard Bond could have published a paper that said, 'Have we found Population III?', to which the answer would have been 'no'. Instead he entitled his paper '*Where* is Population III?',[13] which doesn't demand either a yes or a no, just a frustrated bewilderment. The search for metal-free stars comprised Bond's PhD thesis. Over years he had searched more than half the sky for any stars with less than a 1,000th of solar metallicity, the definition of a Population III star at the time. His survey achieved a lot, and found some pretty interesting stars down to 1/500th of the solar metallicity and lower ... but not many. And none beyond that Population III boundary. I spoke to Professor Bond recently and asked him how much of that paper title came from exasperation:

> I had searched like half the sky ... and I finally just wrote a paper pointing out that there really are very few if any stars with [low metallicity] ... You know, I was just disappointed I put all this effort into finding these things and, and they're just, like I say, needles in a haystack. So, you know, I was just trying to say, 'How can we understand this?' ... Probably when I started out as a naive student, I thought these stars are out there and just look using the right technique and we'll find them, and so here I go to actually look, and where are these dang things? They're not out there ... I was going to find all these wonderful stars ... and they're not out there.

Bond told me that the title of his paper, 'Where is Population III?', was a direct quote from a conference he had attended. An attendee there, American astrophysicist Ivan King, had given a presentation called 'Scientific Scandals', of which the mystery of the missing first stars was one. I love the term 'scientific scandals', as if the Universe were engaged in a cover-up. In a way, though, it is a game where we have been forced into the role of detective. With astronomy, it's often not that the answers are hiding particularly well – after all stars and galaxies are not particularly small. Instead it's that the answer is in plain sight, among vast reams of data, and it is our job to tease out the few lines of numbers, the anomaly in the code, that gives away the position.

As of 2020, the search for the first stars goes on and it goes on with fervour. Bond's methods were sound but limited, able only to probe down to a certain brightness and therefore a certain distance away. We now have methods that can dig much deeper, even out to other galaxies. As well as searching for these stars directly, we also have methods to track their footprints, detecting their effect on their environments and deducing their characteristics. Population III stars are the rare Beanie Baby missing from our shelves, or the mint-condition Penny Black gap in our stamp collection. To find them, we need fresh ideas, resources and stamina.

We'd best start at the beginning.

CHAPTER THREE

The Small Bang

Our story begins with a pigeon, or two. We treat pigeons as a nuisance in urban areas, and they are the number one pest bird in the United States. There are about 400 million pigeons in the world – I can see eight outside my window right now. Where I come from, we refer to them as rats with wings, but they're not as stupid as you may think. There have been a surprising number of studies disproving the notion that pigeons are 'bird-brained'. There are indications that they can discriminate the work of different artists,[1] recognise words,[2] count to nine,[3] and even differentiate between malignant and cancerous tumours in radiology images.[4] One of the most-commonly known positive attributes of pigeons is their homing ability, which can lead to pigeons racing their way home at speeds of up to 97km/h (60mph) over distances of up to 1,609km (1,000 miles). In 2019 a racing pigeon named Armando was sold for more than a million pounds,[5] so celebrated are pigeons' racing abilities. These homing abilities have been selectively bred and honed for millennia because for a long time before Twitter and mobile phones, pigeon post was the fastest way to convey a message. The results of the first Olympics, and news of Julius Caesar's conquest of Gaul and Napoleon's defeat at Waterloo, were all reportedly conveyed by pigeon post.[6] More recently, important war

intelligence pigeon-post messages saved lives across
Europe during the First World War and Second World
War, though not without sacrifices by the birds. Twenty-
thousand army pigeons lost their lives in those wars[6].
Cher Ami, 'Dear Friend' in French, flew 40km (25 miles)
in 1918, despite being blinded, shot in the breast and
having one leg hanging on by a thread. Two pigeons had
been shot down before him but his delivery of the
message 'We are along the road paralell [sic] 276.4. Our
artillery is dropping a barrage directly on us. For heaven's
sake stop it.' stopped a friendly fire barrage and helped
save the lives of 194 men.[7] Thirty-two wartime pigeons,
brave souls such as G. I. Joe, have been awarded the
Dickin Medal, the most prestigious award available to
military-serving animals. At least one, Mary of Exeter,
even has her own little tiny memorial.[8] My point?
Anywhere you look in history you can usually find a
pigeon nearby.

White dielectric material

In 1964, American astronomers Robert Woodrow
Wilson and Arno Allan Penzias were wondering what to
do about the two pigeons nesting in their 6m (20ft)
square horn antenna. The antenna at the Bell Telephone
Laboratory in New Jersey had been built five years
before as a communication conduit to satellites. They
designed the strange sideways ice-cream cornet shape as
a barrier to any signal emanating or bouncing off the
ground, ensuring a clear communication channel with
the satellite. Radio waves are just the right wavelength to
reflect off the Earth's ionosphere, an upper atmospheric
layer of charged particles. This means that we can

purposefully bounce them off, targeting them at an antenna further around the Earth's surface. That's great for anyone listening to Radio 4 in the Scottish Highlands. It's a disadvantage for astronomers, though, because the latest episode of *The Archers* ends up covering our desired astronomical signal. The restricted focus of the antenna on the sky made the Holmdel Horn Antenna an excellent device for taking astronomical observations of the sky, without ground-based interference.

Penzias and Wilson had decided to co-opt the communications antenna as a radio telescope to build on Wilson's PhD thesis, which theorised that there was a faint halo of gas around the Milky Way. He hadn't had the equipment able to measure such a faint background in the face of all the unhelpful terrestrial signals … until now. They made their measurements at a frequency of 1,420MHz because, if there was a halo of hydrogen, it would be sure to emit radiation at this wavelength. The previous users of the Holmdel Horn Antenna had left it set to receive a frequency of 4,080MHz. Coincidentally, the theorised Milky Way halo would be invisible at those frequencies. This was a serendipitous wavelength at which to null test their telescope: turn it on at a frequency where it is not expected to receive any signal and make sure the reading is zero to check the equipment. In June 1964 they turned the telescope to the skies, accounted for known sources of signal within the instrument and the sky, and expected to find a value of zero. In Wilson's words: 'One of the things we wanted to do was to make a measurement of the minimum brightness of the sky … and that went horribly wrong'.[9]

Instead of the null reading they had expected, Penzias and Wilson measured a value of 3.5 degrees above absolute zero, making their measurement 3.5 Kelvin. To rule out a human-made origin, they pointed the telescope to the biggest concentration of humans nearby, in New York City. If this signal was a by-product of televisions, radio or other human undertaking, then a higher concentration of that activity would result in a higher signal in the telescope. Alas, the signal remained the same, at 3.5K regardless of the direction the telescope was pointed. The scientists checked the inner workings of the telescope for the source of the annoying signal, and still the signal remained – it was coming from everywhere in the sky. Sherlock Holmes famously said 'When you have eliminated the impossible, whatever remains, however improbable, must be the truth.'[10] Penzias and Wilson had tried everything obvious, so it's little wonder that they then began to consider the improbable. Two pigeons had made their home right in the depths of the horn of the antenna, beside the heated observation box. They had coated the inside of the antenna in what Penzias politely referred to as a 'white dielectric material'. Pigeon excrement. A dielectric material is an insulating material, a poor conductor that could create interference in the radio signal. These birds had proved themselves a tenacious pair as every time Penzias and Wilson searched a new area of the sky, the whole antenna rotated, toppling the pigeons about as if on a funfair ride.[11] It can't have been a comfortable place to roost, but pigeons do reportedly mate for life and the telescope was their home.

The birds were trapped in a box and, according to Wilson, shipped to a pigeon-fancier. The pigeon-fancier looked at them, determined that they were junk pigeons, and let them go.[9] No sooner had the white dielectric material been cleaned from the antenna, than the pigeons returned to their spring-cleaned love nest two days later. That homing instinct really is something, isn't it? Penzias tells best what happened next:

> We finally – and I'm not happy about this either, but to get rid of them, we finally found the most humane thing was to get a shotgun. We got a shotgun and, at very close range, just killed them instantly. So it's not something I'm happy about, but that seemed like the right thing. It seemed like the only way out of our dilemma.[12]

The role of these pigeons in this story is so legendary you can even visit the exact trap used at the National Air and Space Museum in Washington DC. I'd known the first part of this story since undergraduate days, and when researching this book I spent time finding out the fate of the pigeons in the hope of a happy ending. At least it was quick.* Penzias and Wilson cleaned the telescope, turned it on and now, free of pigeons, they found ... 3.5K. Again. Over the year they had confirmed that the signal stayed the same through daily and seasonal cycles. They had ruled out signals from humans, faulty equipment and pigeon poo. So what was the signal? It turned out that while Penzias and Wilson had thought that the small bang of a shotgun would solve their

---

* The death, not the research.

problems, they had in fact detected the afterglow of the Big Bang.

The Big Bang

The Big Bang is the unembellished idea that all matter in the Universe was squashed into an infinitely dense point until space-time exploded outwards, expanding the Universe we experience today. I may be able to condense it down to one sentence but it isn't an idea that the human brain can easily accept. There was fierce opposition from respected scientific minds of the time, and no wonder. For me, and I suspect many other scientists, the Big Bang is both a preposterous and thoroughly convincing theory. It is a theory that has been forced on our uncomprehending three-dimensional brains, incapable of visualising infinity but able to understand the overwhelming evidence. Even Einstein struggled with the idea that our Universe might be anything other than static. He famously changed his general relativity equations to include an extra term, a 'cosmological constant', to arrest the expansion that naturally fell out of his equations, and keep everything still.[13] Don't feel bad if the idea of a Big Bang is ... ridiculous or even meaningless to you. It was to Einstein, so you can let yourself off the hook. So why is the Big Bang of consequence to the story of the first stars?

As I sit in my office on Earth, staring out at the sky, there is a reasonable question to ask – was there really a *first* star? The Universe around me looks orderly and unchanging, after all. Now and then we might enjoy a solar eclipse, and there will be days when Venus appears brighter and sometimes Halley's Comet comes back around to catch up on things. But over a human lifetime, the Universe will appear much the same at the start as at the end. It's a bit like that summer you had when you were 14. You know the one? Every day you had a bag of snacks, a book and money for sweets, and your friends had their bikes ready to go at the gate. You would toss bread to the pigeons and it was going to last forever. In a Universe full of unimaginable timescales, us wee humans are in our summer holiday once again. It's natural to believe that the Universe today is the Universe as it always was and always will be – a *static, unevolving* Universe. In that case there have always been stars and the question of when the first star came into existence is nonsensical.

Redshifting and blueshifting of light
When you watch an explosion in an action film, there are lots of effects. The debris flies outwards, and the characters feel the heat from the explosion and shy away from the bright light. On a calm Earth day these effects don't tally with the safe, predictable Universe I know, but that's because I am comparing the wrong timescales. I am not viewing the explosion from start to finish and instead I am living one frame of that film in my lifetime. Things don't seem to change because in 80 odd years the expansion is not noticeable. But it is there, and we can measure it. If we look at an image of the Andromeda galaxy, apart from the odd star ending its life in a bright explosion, it will look mostly the same if you took the picture two days or two years later. A person can have a different perception of how fast something moves depending on the context. When watching aircraft take

off in an airport they appear fast. When watching a distant aircraft cross the horizon that speed is much less noticeable. Galaxies are too distant and too large for us to appreciate their movements, but move they do.

Every star comes with its own unique barcode, or spectrum. The light that is emitted by a star covers a broad range of wavelengths, and the gaps in that spectrum, absorption lines, show the presence of certain chemical elements in the atmospheres of the star. The same is true in galaxies but they also contain a lot of interstellar gas and dust, which absorbs wavelengths of light too. Say a galaxy contains mostly Population III stars and just a smattering of the younger Population I and II stars containing heavier elements such as carbon. That galaxy's spectrum would contain virtually no metal lines, and any metal lines that were present would be faint. A younger galaxy will have plenty of metal lines because of the larger proportion of Population I and II stars with heavier elements in them, and the absorption lines will correspond to the same lines that we can detect from irradiated elements in our laboratory on Earth.

In the spectrum of Andromeda, there are distinct absorption lines that match the element calcium. The gaps in the Andromeda spectrum indicate that calcium is present in significant amounts in the one trillion stars in Andromeda, which is not so interesting in itself. The spectrum from Andromeda should contain these calcium absorption lines at the same wavelengths as produced by irradiated calcium on Earth. After all, calcium in one galaxy is the same as calcium in another galaxy. The energy levels present in atomic calcium do not differ depending on location of the atom, so neither should the wavelengths of light absorbed. When we look at the spectrum of Andromeda, however, the calcium lines have moved from where we expect them to be by a few billionths of a metre in wavelength. Because the absorption lines are shifted to shorter wavelengths with respect to their earthly positions, the bluer end of the electromagnetic spectrum, we say that they have been *blueshifted*. It's as if the calcium in Andromeda absorbs photons of slightly different wavelengths, and that the electrons in a calcium atom are arranged differently there, but why would chemical elements vary like that? Well, it isn't the laws of chemistry that are changing but how we perceive the light from the galaxy.

Imagine that we were to light a fire. Pile on all you want, that sequined T-shirt you thought looked cool, a stereotypical box of an ex-partners' junk, the first five drafts of your book … Oh dear, our fire has got out of

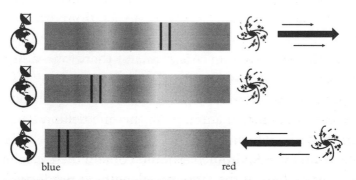

blue   red

Figure 10 The Doppler effect in light waves. The middle row shows a simplified spectrum with absorption lines as observed in a laboratory. Andromeda's spectrum is blueshifted because it is moving towards us (bottom row). Most other galaxies are moving away from us so their spectra are redshifted (top row).

control! As the fire engine makes its way to our smouldering pile of bitterness, listen carefully. When the fire engine approaches us, each successive sound wave is emitted from a slightly closer location to us. Because the fire engine is moving towards us, the emitted sound waves take less and less time to reach our ears, increasing the frequency at which they arrive to us, and we hear a higher pitched sound. After the fire engine leaves us, embarrassed and ashen faced, the time between successive sound waves reaching our ears is increased. The perceived frequency of the sound waves drops and we hear a lower pitch siren. We call this phenomenon the Doppler effect. As a fire engine approaches or recedes, the siren sound it emits never changes, but how we perceive it does. Andromeda is acting like a fire engine, but using light instead of sound. If a source of light is moving towards us, the wavefronts pile up next to each other and we measure a smaller wavelength, a bluer light: blueshift. Conversely, if a light source is moving away from us, then the wavefronts are more dispersed, and we measure a lower frequency, red light: redshift. For Andromeda the calcium absorption lines appear at bluer wavelengths than those we know that are produced by the element, suggesting that Andromeda is moving towards us.

With the fire engine, the faster it goes, the more pronounced the change in pitch will be, as it approaches a small fraction of the speed of sound, over 340m/s. Light travels at just under 299,792,458m/s. This explains why we don't notice the fire engine's spinning light changing colour as it moves towards or away from us. The fire engine isn't fast enough to catch up with the light waves in any appreciable way, so while the shift in wavelength happens, it is miniscule. For Andromeda, calculations show that the tiny shift in the wavelength of those calcium absorption lines indicate that the gigantic spiral galaxy is on a collision course with our own Milky Way at a speed of approximately 110km/s.[14] In a few billion years, our galaxies will collide, but it's interesting to point out that the collision will not be a violent explosion as you might imagine. Stars are spaced so far apart in galaxies (on average 50 trillion kilometres apart) that the probability of any two stars colliding is very low. The gravitational distortion of the two galaxies will be profound however, and they will blend to form a gigantic elliptical galaxy sometimes referred to as Milkomeda (terrible name).

The expansion of the Universe

We have established that when we observe galaxies we can find out if they are moving towards or away from us. When we make similar measurements of the galaxies all around us, the vast majority of other galaxies' spectra are *redshifted*, meaning that the galaxies are moving away from us. They are embedded in a space which is expanding all the time. Andromeda is an exception to the rule – the strong local gravitational attraction between the two galaxies is pulling them together. Our Universe appears not to be full of galaxies wandering randomly like pigeons in Trafalgar Square, but a one-way thoroughfare of racing pigeons. The pattern is the same whatever direction you face: the vast majority of galaxies are flying away from Earth, like the debris from some big explosion. There is something else interesting about this redshifting. The further away a galaxy, the higher the speed at which it is travelling from us. Belgian astronomer Georges Lemaître and American astronomer Edwin Hubble were the first to acknowledge this relationship in the late 1920s.[15, 16] The Hubble–Lemaître law states that the speed at which the galaxies are moving away depends on their distance from us: the further the galaxy, the faster it moves.

When we first learn that everything is moving away from us, it is easy to conclude that this means that we are the centre of the Universe. To see why this idea is wrong we can imagine covering a deflated balloon with stickers. As we inflate the balloon, the stickers move apart. Whichever sticker you follow, all the other stickers appear to move away from it, just like the galaxies we observe. Every galaxy in the Universe will observe the majority of other galaxies moving away from itself. This illusion of centrality is therefore shared by all galaxies, and of course cannot be true for all of them. If you follow any one sticker, you will notice that the nearby stickers only appear to move a short distance away. Stickers further away to begin with have travelled much further because there was more expansion of the balloon in between. The nearby and farther stickers cover this short or long distance over the same time, and so appear from our sticker to be travelling at different speeds. The increasing speed with increasing distance is the hallmark of how objects behave in an expansion. This *Hubble flow* of galaxies away from Earth was one of the first pieces of evidence that our Universe is expanding. If we rewind our action–film sequence, a natural conclusion to everything moving apart is that, in the past, everything was very close together. The evidence pointed to a Big Bang. There was a second camp of astronomers who weren't convinced. They found an evolving Universe distasteful, even if they could accept that it was expanding. The Steady State astronomers believed that as the Universe expanded new matter filled in the gaps. This dispensed with the idea that matter flying apart now meant that it had been much closer together in an infinitely dense point. The Steady State astronomers couldn't tell anyone where this extra matter came from, but the Big Bang astronomers couldn't tell anyone why our Universe began as an infinitely dense point. Both camps were plausible at the time.

The first nail in the coffin for Steady State theory came when English radio astronomer Martin Ryle made a survey of galaxies in the sky. In a Big Bang Universe, galaxies closer to the present (in distance and time) were more spaced out because there had been more expansion. This means that there should be more galaxies in a given field of view the further away you look. In a Steady State universe the number of galaxies would remain the same independent of how far you look back in time or space because the theory is of an unevolving Universe. The further Ryle looked in distance, the further he was looking back in time, and the more galaxies he counted[17]. A resounding victory for the Big Bang theory. Considering the stakes and the level of commitment in both camps this wasn't enough to end the argument. The argument between Steady State scientists and Big Bang scientists eventually ended only because of the second hallmark of an explosion: the afterglow.

The cosmic microwave background

While Penzias and Wilson were trying to figure out the mysterious signal, four astronomers were setting up their own radio antenna, 60km (37 miles) down the road at Princeton University. They had the express aim of detecting a background radiation residual from the Big Bang. P. James Peebles, Robert Dicke, Peter Roll and David Wilkinson had realised that the early Universe would be extremely hot after the Big Bang, just like any explosion. The very early Universe was full of electrons, protons, neutrons and photons all at the same temperature, in thermal equilibrium. An object in thermal equilibrium emits light of wavelengths across the electromagnetic spectrum at intensities dependent only on the temperature of that object. If there was a Big Bang then there should be leftover radiation with the characteristic spectral shape of a blackbody, the shape we saw in Figure 5. The first step would be to find a signal at all, at any wavelength, which was within the predicted temperature range. Roll and Wilkinson had even got as far as building the antenna when they heard about Penzias and Wilson's detection. The Princeton team visited Penzias and Wilson to check the equipment and data. Neither Peebles, Dicke, Roll or Wilkinson could find any fault with the set-up, and in Dicke's words at the time 'Well boys, we've been scooped!' According to Peebles, Penzias exclaimed 'Well, that's a big relief. We understand this thing at last. Now we can forget it and go and do some real science!'[11]

They announced the discovery in a pair of research papers published in 1965, one from the Princeton theory team,[18] and one from Penzias and Wilson.[19] The Penzias and Wilson paper is noticeably careful in its wording, sticking to the experimental outcome and taking up only a page of text. It was a discovery of unparalleled importance for our understanding of cosmology, and for that Penzias and Wilson won the 1978 Nobel Prize for Physics.[20] In October 2019, Peebles also won the Nobel Prize 'for theoretical discoveries in physical cosmology', to which his prediction of the CMB contributed a large part.[21] They had detected what we now know as the cosmic microwave background (CMB), the radiation left over from the Big Bang. The Universe started from a Big Bang. The Universe began. Time started. There was a first of everything, including stars.

The Penzias and Wilson measurement constituted one data point on the expected blackbody spectrum of the cosmic microwave background. Attention turned to filling in the rest of the wavelengths. Penzias and Wilson's experiment was ultimately limited due to it being ground based. They could only make measurements at very few frequencies, not enough to uncover the spectrum of a blackbody. Even though photons are able to traverse through gas and between galaxies without being deterred from their path, when it comes to the Earth they practically stop dead. This is because the energy of a CMB photon now is very low – low enough that it is easily absorbed by water in the Earth's atmosphere. To really investigate the CMB, we have to get past the Earth's atmosphere and head into space. Three missions have done this so far: the Cosmic Background Explorer (COBE, launched in 1989), the Wilkinson Microwave Anisotropy Probe (WMAP, 2001) and Planck (2009), taking measurements at multiple frequencies and confirming that the CMB is indeed a blackbody spectrum down to at least one part in 10,000. The small deviations from the spectrum are the result of tiny temperature fluctuations across the sky, which WMAP and Planck captured in exquisite detail. These fluctuations encode information about the structure of our Universe at the point at which the radiation was produced, 380,000 years after the Big Bang. The data also contains clues as to the state of the Universe since, which will be key for our later search for the footprints of the first stars in Chapter 10.

The early Universe: inflation
We can now write a narrative about how our Universe began and set the scene for those first stars – 13.8 billion years ago the Universe, space-time itself, began expanding from an infinitely dense point. At first this expansion was exponential, during a period cosmologists call *inflation*. We can visualise the impact of exponential growth using an Indian origin story of the game chess.[22] In the version I know, a selfish and bored king enjoys feasts and fineries while the people of the land starve, giving all the rice they grow in taxes. One day a peasant approaches the King and says that he has invented a game for him, called chess. The two play and play, and in his gratitude the King asks the peasant what he would like as a gift. The peasant replies that he would like the King to place a single rice grain on one corner square of the chessboard. Then, he should place two grains in the second square, four in the third, eight in the fourth and so on … doubling the amount of rice each time. Laughing off the

request, the King shouts for his servants to bring him a bowl of rice grain. He begins the task. By the end of the first row he has placed 255 grains and shouts for a sack of grain. By the end of the second row he has placed 65,535 grains. He calls for more sacks and continues, emptying his kitchen and stores of all the rice grains. Eventually he runs out of the rice grains, and calls for his servants to gather more rice from the land. The peasant explains that there is no rice in all the land, for the people are starving. His intent was only to make the King notice the famine, and all he wished was for the King to stop taking the rice from the people. The King agrees, chastened, and the tale ends, hopefully without any stray grains being eaten by a pigeon because they would explode in a little bang.* And a good thing, really. If the King had finished filling that chessboard he would have used 18,446,744,073,709,551,615 grains of rice. The number of grains per square increased in what we call exponential growth, the same growth that the Universe underwent after the Big Bang. Only a tiny fraction of a second after the Big Bang the early Universe expanded by a factor of at least 100 trillion trillion, before settling into a more relaxed expansion. Inflation is very much an open and active research field but, together with the Big Bang, forms what we accept as the standard cosmological model.

The early Universe: nucleosynthesis
I started out this section asking 'Is there a first star?' The Big Bang theory tells us that there was a first of

---

* Not true. Old wives' tale.

everything, because the Universe itself had a beginning. We have touched on topics such as the inflation of the Universe and infinitely dense singularities, so no question seems too fantastical at this point. Did the first star just pop out of the Big Bang fully formed? Are the galaxies literal debris of a gigantic explosion, somehow minimised to fit into a singularity but inflated to their full size today?

A few seconds after the Big Bang, the Universe largely contained radiation and the basic building blocks of all you see around you: subatomic particles such as protons and neutrons. At these early times, the contents of the Universe were at the same temperature in thermal equilibrium. The making and breaking of atoms averaged out and everything stayed about the same in terms of its proportion. Imagine a playschool room full of toddlers and Lego. Half the children might be intent on building towers, while half might be intent on running about and smashing towers. If all the kids are hyper, then as soon as a child builds a set of bricks into a tower, it is broken up again. The pieces may find themselves part of other towers … before being smashed up again. At any one time, there will be a certain number of towers in existence, but any one tower does not last very long. In thermal equilibrium the number of bricks (protons and neutrons) never changes, the number of charging toddlers (radiation) never changes, and while the number of towers might fluctuate in any given second, it is too chaotic for any tower (atom) to exist for very long at all. It is certainly too hot for any larger structures to be created out of atoms, such as stars. We are used to talking on such grand scales in terms of distance and time, that it can be quite a shock to talk about the pace of the very early Universe. The first three minutes of the Universe were so packed full of events that they have merited their own book of the same title, *The First Three Minutes* by Steven Weinberg.[23] About 13.82 seconds after the Big Bang the Universe hit a sugar crash, as it expanded and cooled down enough to allow nuclei to build up. Nuclei are structures comprising the subatomic particles protons and neutrons. A nucleus together with at least one bound electron constitutes an atom, but in the very early Universe it was too hot for the electrons to bind with the nuclei. In our anecdote, the charging toddlers have tired and the towers can get a little larger. Now, the larger the tower, the less stable it is, and I can say the same with low-mass nuclei, though this analogy does not extend across the periodic table. While we can build up hydrogen and helium nuclei, building up anything larger is still very difficult. Even a low-energy photon can break up volatile molecules heavier than hydrogen, just as even a lethargic toddler can break up a teetering, tall tower while wandering past it. When nucleosynthesis (the term for making nuclei) ended 3 minutes 46 seconds after the Big Bang, about 25 per cent of the normal mass of the Universe had settled as helium-4, whose nucleus is formed of two protons and two neutrons (we'll get on to the abnormal mass in Chapter 4). The rest of the mass, 75 per cent, was left as hydrogen nuclei (protons), with around 0.01 per cent deuterium, helium-3 (a flavour of helium made of two protons and one neutron) and a smidgen of lithium. By the time the Universe had time to poach an egg, it had finished its hard work and created all the hydrogen and helium that was needed to create the first stars a hundred million years or so later.

After that four minutes of hard work, the Universe remained a mosh pit of nuclei, electrons and photons, too chaotic to allow photons a clear path anywhere. Photons were colliding with electrons everywhere and their pathways looked like a staggering drunk's might. If you have ever driven along a road on a very hot day, you might have spotted a highway mirage: a wavering of the air above the road. The varying temperature of the air above the road results in a differing amount of refraction in the air, deflecting the photons from their straight paths and forming the optical illusion of a wavering or missing road. We cannot see anything with light before 380,000 years because the environment is too volatile to allow photons to travel on unimpeded paths to our telescopes. At this point, about 380,000 years after the Big Bang, the free electrons bound to the free protons, creating hydrogen atoms in an event called *recombination*. With the fall in free electrons, the photons found themselves able to travel freely for the first time, and we say that the Universe had become transparent to radiation. Together, the spectrum of the photons followed a blackbody distribution. It is this radiation that Penzias and Wilson measured with the Holmdel Horn Antenna. This cosmic microwave background radiation is everywhere because it pervaded every part of the tiny, hot Universe when it exploded from a singularity. The Universe may have got a lot bigger but the radiation, the afterglow, still pervades it. The photons we measure today still retain that blackbody spectrum, the exact distribution we would expect if the Universe began in a Big Bang. The CMB is there, just as it was in 1964, and continues to be picked up as interference by radio antennas. If you remember the days of analogue television, a decent amount of the static on untuned channels was radiation from the Big Bang.

★ ★ ★

This chapter set out to justify the theory that there are indeed first stars to look for. We have cited two major pieces of evidence for the Big Bang theory: the recessive motion of the galaxies around us and the detection of the cosmic microwave background. Using the known absorption lines produced by certain chemical elements, we can measure the speed and direction of movement of the galaxies around us. We observe their light as redshifted almost without exception, indicating that the galaxies are moving away, giving away the underlying expansion of the Universe. Over decades we have also pinned down the relic radiation from that Big Bang, so that we can reproduce the predicted blackbody spectrum and, even more, map out the tiny temperature deviations across the sky. This evidence has positioned the Big Bang theory as practically irrefutable in the minds of most scientists. We can use this theory to make conclusions about the state of the early Universe, namely that it was too hot and chaotic for anything much heavier than helium to form. The first stars did not pop out fully formed from the Big Bang, and instead would have to form later from very few ingredients, only hydrogen and helium.

After becoming transparent to radiation, the Universe settled into a very long period of rest. Anyone sitting in the early Universe would have found it boring, dark and uneventful to the human eye, as everything just continued to cool down as the Universe expanded. It would be a hundred million years or more before there would be something interesting to look at.

CHAPTER FOUR

A Lucky Cloud of Gas

Deep down in the southern isles of Japan lies Naha, the prefecture capital of Okinawa. Every year in the 'land of the rising Sun' a highway is closed to accommodate the unfurling of a 183m (600ft) long rope, several feet in diameter and weighing as much as 30 cars.[1] Smaller side ropes, woven into the larger rope, unfurl across the road at a 90 degree angle and 15,000 participants grab on. The tug-of-war between East and West teams is a celebratory event where participants and onlookers wish for happiness and peace, prosperous business and good health. As the gong sounds, a fierce struggle begins: the team that pulls the rope in its direction by 5m (16ft) within half an hour, wins. As of October 2019, the East team has come out top in the event's history, with 16 wins, 14 losses and 16 draws.[2] Anyone can grab a rope as long as they get there early enough (about 280,000 people attend in all), so this is an excellent exercise in statistics as well as strength. If these were trained teams the results would be more stochastic, but having teams of 7,500 composed of random women, men and children, the teams are much more likely to be evenly matched. The crowd averages out the stamina of the odd highly trained athlete or the strength of the Okinawan military base personnel. You would expect to see results much more in line with equally matched teams and so far, that's what the 16-16-14 East–Draw–West results nicely portray.

When I think about stellar formation, I can't help but think of a game of tug-of-war, on an astronomical scale. Photons were free to move about 380,000 thousand years after the Big Bang, and they formed the cosmic microwave background, as we have read in the last chapter. The Universe had settled down from its early tumultuous years and was a steadily expanding pantry of light elements. Across the Universe these ingredients formed clouds of various sizes and temperatures. To form a star, the conditions in a cloud of gas must be just right. It is a struggle so closely matched in strength that the event can go on for billions of years. Declaring a draw is not an option in this case, though. One team, gravity, has infinite stamina, and while the opposing pressure produced by internal nuclear reactions provides relief, the nuclear fuel will eventually run out. The star will falter while gravity pushes on relentlessly. We might define a star in the driest of terms: a celestial body where the pressures created by internal nuclear reactions are enough to hold up its own collapse due to gravity. Really, though, a star is just a very lucky cloud of gas.

Powering the stars: gravitational contraction
Let's introduce our first contestant. Gravity is one of the four fundamental forces in the Universe. The others are the weak and strong forces that govern nuclear reactions and the electromagnetic force, which acts between charged particles. Gravity is an attractive force that exists between any two objects with mass. The more massive

an object, the greater the gravitational force it will exert on other objects. In everyday use we tend to use the word 'massive' to mean something enormous in volume, but scientifically it relates to the amount of matter making up an object where the size of the object is neither here nor there. For example, a marble is more massive than a ping-pong ball, despite being smaller in size. A gas cloud in the early Universe was a loose collection of hydrogen atoms, perhaps with the odd heavier atom such as helium in the mix. If we follow a hydrogen atom in our cloud, it feels a tiny gravitational pull from every other atom in the cloud. Those gravitational forces will cancel out so that the atom feels an overall force towards the centre of the mass of the cloud. If our atom is moving at low speed, it has what we call low *kinetic energy*. It doesn't have the energy to counteract that gravitational pull, in the same way that I don't have the energy to counteract the gravitational pull of the Earth and jump to the Moon. The atom moves closer to the centre of the cloud. An atom with a large kinetic energy does have enough energy to counteract that gravitational pull and can continue on its way and escape the cloud. If most atoms cannot counteract the overall gravitational pull of the cloud, we say the cloud is a gravitationally bound system. Conversely, if most of the atoms have plenty of kinetic energy and can escape that gravitational pull, we say the cloud is unbound. It is unlikely to exist as a cloud for much longer.

We can understand the effects of gravity further by thinking in terms of acceleration and energy. When a force is applied to a static object, it moves. It changes

velocity or *accelerates*. For example, the acceleration
because of Earth's gravity is 9.81m/s². This means that,
in the absence of complicating factors such as wind
resistance, an object falling from a height above the
Earth's surface would accelerate towards the ground
gaining 9.81 metres per second in speed, every second.
The larger the acceleration experienced by an object,
the greater the force it is being subjected to. Let's
imagine a whale suddenly appearing several miles above
the Earth as Douglas Adams did in his book *The
Hitchhiker's Guide to the Galaxy*.[3] 'This is not a naturally
tenable position for a whale, this poor innocent creature
had very little time to come to terms with its identity as
a whale before it then had to come to terms with not
being a whale any more.' A stationary whale appearing
1km (0.62 miles) above the Earth would find itself
subject to a gravitational force from Earth and it would
accelerate towards the ground, encountering it at a
speed of approximately 140m/s, or 310mph. If we
performed the same experiment above the far more
massive Sun (new whale), where the gravitational pull is
about 274m/s², then the whale would meet a fiery doom
at a considerably faster speed of 740m/s. Though, to be
fair, there wouldn't be a ground to hit, more a point of
vaporisation … but either way it would be a quick way
to go. Gravity accelerates objects equally so that, without
wind resistance, if you drop a hammer and a feather at
the same time they will hit the floor at the same time.
This is difficult to do on Earth as the wind resistance
conspires with the feather's greater surface area to slow it
down compared to the hammer. NASA, always ready
for a laugh, performed this experiment on the Moon in

1971.[4] Watching the video you can see that the feather
and the hammer do indeed fall to the surface of the
Moon at the same time.*

We can see the link between force and acceleration ...
but what about energy? When a force results in movement the energy of the object changes. The object might
gain kinetic (moving) energy and move faster; it might
gain height, gaining *gravitational potential energy*. It might
fight against a friction force, losing kinetic energy and
gaining thermal energy (energy created when atoms and
molecules move faster due to a temperature rise) as its
contact surface heats up. The 15,000 participants of the
Naha giant tug-of-war have to spend an enormous
amount of energy pulling on the rope. They will lose
some stored chemical energy from breakfast as thermal
energy, as their muscles heat up from the effort. Some
energy will be transferred into the rope, deforming it

---

* Astronauts have to undergo years of training, on top of illustrious
careers often in the military forces. Some of them still keep a sense
of humour, though, and surprise the people at home. Alan Shepard
was the first American to travel into space and in 1971 he became
the only person to play golf on the Moon. Using a club head
attached to one of the scientific sample-collecting tools, he hit two
golf balls a short distance. While the Moon's gravity is low enough
for golf on the Moon to be a much longer range sport, the range of
movement allowed by his spacesuit was small, resulting in the shots
being more like chips. On the same mission, *Apollo 14*, lunar
module pilot Edgar Mitchell performed his own experiment, in
secrecy from his commander, Shepard, and mission control. Mitchell
transmitted mental images of random shapes to four accomplices
on Earth, to prove psychic powers. It didn't work, not least because
those on Earth forgot to adjust the pre-arranged times for the delay
in the launch.

and charging it with a small amount of elastic energy. When it comes to a tug-of-war, failing to consider this last energy transfer can be a fatal mistake. Elastic bands are 'elastic' because they have a large potential to store elastic energy: when they are deformed, they will use all that stored energy to spring back into position. When improper rope is used for a tug-of-war, a similar rebound can occur. Nylon rope, for example, can store a significant amount of elastic energy. If subjected to enough stress, for example during a tug-of-war, the rope will release all that energy as it snaps back. In 1997 in Taiwan, 1,600 participants pulled a 5cm (2in) thick nylon rope. Inadequate to the task, the rope broke and recoiled, ripping an arm each off two participants.[5] The Naha rope is made of rice straw and is woven to such a degree that, while the rope actually broke during play for the first time in 2019, no injuries were reported and the contest was a draw.[2]

In the Universe energy must be conserved: we cannot create or destroy energy, only transfer it. The law of energy conservation is essential for us to understand how a star not only forms but survives and eventually dies. It also explains our whale's high-speed doom. High above the planet, the whale had a whole lot of what we call gravitational potential energy. The amount of energy it would take to get a whale a kilometre above the Earth, fighting against gravity, is quite large. Think how much energy you need to jump a centimetre off the ground. Now consider how much energy you need to jump 20cm (8in) off the ground. If you wanted to jump 2m (6.5ft) off the ground, you would need an external injection of energy, for example from someone else using their stored energy to lift you as you jumped. All the time you are jumping, you are working against the gravitational pull of the Earth. At the height of your jump, all that chemical energy in your legs has been converted into gravitational potential energy, which then converts into kinetic energy as you fall back to Earth. When our whale appears a kilometre above the Earth, it has a gravitational potential energy in relation to its position above the Earth. As the whale falls, that gravitational potential energy transforms into kinetic energy: the whale moves faster and faster and faster ... splat. The gravitational potential energy that our whale possessed at the start of this experiment will have almost entirely converted into kinetic energy. In an ideal experiment the conversion would be total, but in reality the whale would have encountered air resistance, producing friction with the molecules in the surrounding atmosphere, heating them ever so slightly and converting some of its gravitational potential energy into thermal energy. When you change the energy of one element of a system, for example by increasing the height of an object above the Earth, the energy in order to achieve that has to come from somewhere. Conversely, if you decrease the height of an object, lessening its gravitational potential energy, that energy has to go somewhere – it cannot just disappear.*

As our cloud of hydrogen atoms contracts under the combined gravitational force, its gravitational potential

---

* In my whale thought experiment I have taken some liberties, having the whale simply appear imbued with gravitational potential energy.

energy is lessening. Each atom is getting closer to the centre of the cloud, like a tiny whale hurtling towards the ground. The atoms accelerate towards the centre and rub up against each other much more, creating a lot of thermal energy. The atoms and molecules move about a lot more, their kinetic energy increases and we lose some energy as radiative energy: photons escaping the cloud, transferring light and heat. We are in the protostar stage … the foetal stage of a star. It's the beginnings of a star, but we aren't fully there yet.

But hang on … aren't stars known for their heat and light output? Have we already made a star just by subjecting a cloud of hydrogen to gravity? In the early twentieth century, when astronomers were trying to work out what made the Sun shine, they proposed the idea that it was gravitational contraction alone that accounted for the heat and light from the Sun. If a cloud releases energy in the form of light and heat as it contracts, then is that all there is to a star? The problem with this idea is that the collapse is far too fast. The Sun could only release energy in this way for tens of millions of years before it couldn't get any smaller.[6] This is a long time but far too short to tally with geological evidence showing that the Earth, and therefore the Sun, is billions of years old.

Our tug-of-war is so far pretty one-sided. Gravity is relentlessly pushing down, compressing the hydrogen. For the not-so-lucky gas clouds, the increased pressure as the gas compresses is too great to contend with, and the gas cloud dissipates. Those that can collapse, continue collapsing, releasing heat and light – but this can't be the whole story, otherwise the Sun would have stopped emitting light 10 million years after it was born. The pressure continues to rise at the centre of the cloud until the atoms are energetic enough to ignite a new kind of reaction.

Powering the stars: combustion

Whether we are talking about the early Universe or the present day, the main ingredient of stars is hydrogen. This is the lightest element in existence, consisting in its usual form of only one proton and one electron. Hydrogen is quite an unremarkable element: it is odourless, colourless, tasteless, non-toxic and not even rare. It's easy to come by on Earth, though it is so ready to form bonds with other atoms that it is mostly found as part of larger compounds: it forms the H in H_2O (water), for example. In fact, hydrogen atoms are so reluctant to go it alone that they pair up readily, forming molecular hydrogen, H_2, comprising two hydrogen atoms coupled together.* Even paired up, doubling their mass, molecular hydrogen is very light and so very handy for making things float upwards. However, its usefulness as a buoyancy aid is offset by its characteristic flammability. The unfortunate German passenger airship, the *Hindenburg*, famously went up in flames as it came into land over New Jersey on 6 May 1937.[7] The cause of the disaster is still disputed, but a leading theory is that a spark of static electricity triggered by the anchoring of the descent ropes caused a small hydrogen leak to ignite. This then consumed the

---

* 'Hydrogen' derives from 'water-former' in Greek, named so in 1783 by Antoine Lavoisier, because of the way hydrogen formed water (H_2O) when combined with oxygen.

fabric that made up the thick, rigid skin of the airship. Thirty-six people died, though when you see the images it is quite remarkable that the other 60 passengers survived. The *Hindenburg* was not the first airship to ignite, but it scared people enough to ground commercial airship flights for good, despite a safer gas alternative being available. The airship featured in Chapter 1, the *USS Los Angeles*, arrived in the United States from Germany filled with hydrogen but this was quickly changed upon arrival to helium.[8] Helium has a nucleus comprising two protons and two neutrons. It is difficult to capture on Earth and thus expensive. The larger mass of helium (about four times that of hydrogen) results in it being less buoyant compared to hydrogen. The difference is tiny – helium is approximately 93 per cent as buoyant as hydrogen. Under conservative atmospheric conditions, $1m^3$ ($35ft^3$) of hydrogen could lift about 90g more than the same volume of helium. That might not seem a big difference but when you scale it up to the volume of the *Hindenburg*, it would have meant leaving tens of thousands of pounds in weight of payload (that is, all the passengers) on the ground, making the commercial angle of the venture flawed. Despite the tendency of high concentrations of hydrogen to explode, this flammability is still not enough to explain the energy output of the Sun. After all, to burn something you need oxygen, and oxygen is in low supply in a Universe predominantly made of hydrogen and helium. If we look to one of our main energy sources, coal, we can calculate that if our Sun was a pile of burning coal emitting energy at the rate it is, it would run out of fuel in only 6,300 years.[9] Clearly, the Sun isn't simply on fire.

But hydrogen can do something even more catastrophic than burning – it can fuse.

Powering the stars: nuclear fusion

More than 1,000km (621 miles) to the north of Naha, yet still in Japan, lie two cities: Nagasaki and Hiroshima. Due to the terrifying and traumatic events linked with those locations, you only have to hear the names once to remember them forever. On 6 and 9 August 1945, two nuclear weapons nicknamed Fat Man and Little Boy landed on the cities, killing at least 100,000 people instantly. Over the next months many more would die of radiation sickness, burns and other related injuries. The weapons devastated the cities themselves, flattening them in a radius of 1.6km (1 mile) from the drop site. As the bombs detonated, they released a huge amount of thermal radiation and light across the spectrum. One of the pilots present when the Hiroshima bomb was dropped later said that '... this bright light hit us and the top of that mushroom cloud was the most terrifying, but also the most beautiful, thing you've ever seen in your life. Every color in the rainbow seemed to be coming out of it.'[10] It is hard to fathom that the smallest of things, a mere atom, created this devastating and large-scale release of energy.

Atomic bombs work on the concept of nuclear fission (meaning 'to split') and use fuels comprising *isotopes*. These are atoms of an element that have the same number of protons but different numbers of neutrons. Deuterium is an isotope of hydrogen, its nucleus consisting of one proton and one neutron. Tritium is another isotope of hydrogen and contains one proton and two neutrons. Isotopes of atoms are very common, but it is only hydrogen whose isotopes have different names. We refer to the isotopes of every other element using a number to indicate the total number of protons and neutrons (called the mass number). For example, carbon-12 has six protons and six neutrons, while carbon-14 has six protons and eight neutrons. Nuclear fuel, such as uranium-235 or plutonium-239, is chosen for its inherent instability. If a neutron collides with a uranium-235 atom with enough speed, the atom splits. The products of that are smaller atoms, called fission fragments and, crucially, three more neutrons and a lot of energy. Those neutrons trigger more uranium to split, releasing energy and more neutrons, and so on. There is a chain reaction until the system is overwhelmed and there is a large explosive release of all that energy.

The thermonuclear device, or hydrogen bomb, is the more modern version of the atomic bombs deployed during the Second World War. In a hydrogen bomb there is an extra stage on top of the fission stage, in order to create even more destructive power. Hydrogen bombs ignite a fissile material such as enriched uranium near some *fusion* (meaning to join) fuel such as deuterium and tritium. The energy from the fission reaction compresses and heats the fusion fuel until it fuses into a heavier element: helium. The atoms created in the fusion and fission reactions are smaller in total mass than the original atoms. For both fission and fusion, a loss of mass results in a big release of energy, and Einstein can tell us why. Einstein's most famous equation, in fact probably *the* most famous equation, tells us that energy is equivalent to mass, with a multiplier of the speed of light squared $E=mc^2$. The speed of light we know is a *big* number: 300 million m/s. Even a small loss of mass results in a huge release of energy.

The science of nuclear fusion in stars is the same as in thermonuclear devices, except that stars start out with predominantly hydrogen atoms and must therefore first synthesise deuterium and tritium. The proton–proton chain reaction,[11] is the dominant fusion process in stars like our Sun and also the first stars. In present–day stars larger than about 1.3 solar masses there is also another fusion process facilitated by the metals carbon, nitrogen and oxygen. In lower mass stars the temperatures are not great enough to trigger this process, and in metal-free first stars it was not an option at all, due to the complete lack of these heavier elements.

The fusion of hydrogen into helium results in a net loss of mass, and an equivalent release of energy. The mass of the helium nucleus is about 0.7 per cent lighter than the mass of the four ingredient protons. That's a tiny difference and is equivalent to only 4,000 billionths of a Joule of energy. For reference, a square of chocolate provides you with 100,000 Joules. While a single fusion event releases considerably less energy than a chocolate bar, the sheer amount of hydrogen in the Sun means that every second 600 billion kilograms of hydrogen are turned into 596 billion kilograms of helium. Every second, then, the Sun is converting 4 billion kilograms of mass to energy,[9] and there's plenty left. If we were to convert just 10 per cent of the Sun's hydrogen to helium we would produce enough energy for the Sun to shine for 10 billion years.[6] Hydrogen fusion reactions are

more than enough to explain the energy output of the
Sun. Vitally, the release of the energy in these reactions
maintains extreme temperatures in the core, resulting in
atoms zipping about at high speed and creating a thermal
and radiation pressure that can counteract the push of
the gravity, stopping the collapse. A second contestant
has picked up the rope: gas pressure. We now have two
competing teams in our tug-of-war, though with stars it
is more a push-of-war: gravity pushing inwards and gas
pressure pushing back.

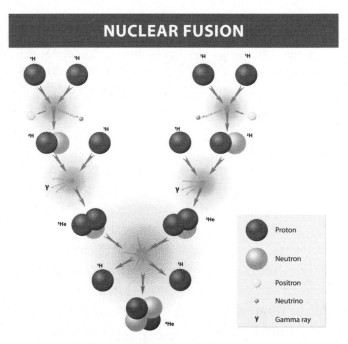

Figure 11 *Stellar fusion. The proton-proton chain reaction is dominant
in modern stars of less than 130 per cent the mass of the Sun and,
because of the lack of metals, the first stars.*

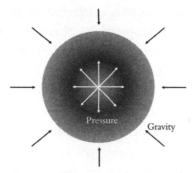

Figure 12 *The stellar tug-of-war. The gravitational force exerted by the mass of the star causes it to contract. The nuclear reactions in the star's core provide energy and, vitally, a pressure that pushes back.*

A stellar bomb?

So stars are gigantic thermonuclear devices. Alarming. On Earth, we can control fission reactions in nuclear reactors to produce energy. This is achieved by the use of neutron poisons – ultra neutron-absorbent materials that ensure only one of the three neutrons released in the fission reaction goes on to produce another reaction. In this way, the chain reaction is controlled. The method is not infallible, however. Japan provides an example of when this control failed, in the Fukushima disaster. On 11 March 2011, a magnitude 9.0 earthquake occurred off the Pacific coast of Japan. The earthquake caused the reactors to shut down, in line with the safety procedures. Even after the reactors have shut down, there is a huge amount of residual heat, which is removed by pumping coolant around the core. Due to problems in the electricity grid following the earthquake, the coolant had to be pumped using the back-up diesel generators. This did not work for long, however, as the tsunami overwhelmed the station, flooding the generators. Without any new coolant the core began to overheat, causing three nuclear meltdowns. The nuclear fuel melted under the increased temperatures, breaching the containment and resulting in a serious leakage of radiation into the sea water and surrounding area. In addition, the temperatures soared so high that the hydrogen was stripped away from the water molecules in the coolant steam, creating high concentrations of hydrogen that then did what high concentrations of flammable hydrogen are very good at doing and exploded. The Fukushima disaster clean-up will be a long and difficult process: the removal of the radioactive molten nuclear fuel is not expected to begin until 2021.[12] Overwhelmingly so, nuclear power is a controlled and safe medium for energy production, and even when disaster strikes it does not result in the kind of runaway uncontainable chain reaction that we ourselves choose to initiate when using nuclear weapons, and that happens naturally within the Sun. Happily for us, the Sun is its own container, since the nuclear reactions are only happening in the very core. The sheer mass of the Sun and the gravitational pressure that the outer layers of the Sun exert on the core of the Sun prevent that huge release of energy from exiting in a fast and uncontrollable fashion – just like on Earth if someone throws themselves on a grenade, they are unlikely to survive, but the resulting explosion will be a lot more contained, potentially saving the lives of everyone else in the immediate area.

Recreating this controlled fusion is a holy grail of nuclear physics. While uranium and other fissile isotopes are rare, hydrogen is cheap and plentiful and nuclear fusion produces less radioactive waste. If we could figure out a way to recreate the incredible gravitational pressure that sparks fusion, we would have a virtually free energy source to sustain humanity. This amounts to recreating the physical conditions in the middle of the star.

The state of stars
For a long time, the Sun shouldn't have been able to exist. Two properties of the Sun were in direct contradiction. The Sun's high temperature indicated that it was most likely a gas. Gases tend to have very low densities, where density is how much mass there is within a set volume. A cube of length 1m (3.28ft) filled with water would weigh 1,000kg (2,204.6lb), giving water a density of $1,000kg/m^3$. A similar cube made of wood would weigh 700kg (1,543lb), giving wood a density of $700kg/m^3$, whereas concrete is closer to $2,400kg/m^3$. Gases have a much lower density – for hydrogen it is $0.09kg/m^3$, at normal Earth pressures and temperatures for example. When we consider the mass and volume of our Sun we find that the average density is $1,400kg/m^3$. This indicates that the Sun is not a gas, yet the temperatures suggested by the solar spectrum are too hot for most materials to remain a solid or liquid. Discovering these two properties of the Sun coexisting was akin to touching an ice cube and finding it to be hotter than steam. And yet, the Sun does stubbornly continue to exist.

When you think about states of matter, you will invariably have been taught that there are three: solid, liquid and gas. You are told solids keep their shape and volume, like ice; liquids take the shape of their container, like water running from a tap; and gases fill the entire volume of a container, as we see with steam. However, there is a fourth state called plasma, and it makes up the state of most of the known matter in our Universe (excluding dark matter, see Chapter 5). Plasmas were not readily observable on Earth for a long time, as they require extremely high temperatures or strong electro-magnetic fields to exist. One common exception is lightning, which is a cloud-to-ground plasma column.

In the Sun, temperatures are so high that negatively charged electrons within atoms are stripped away from the positively charged nuclei, leaving a gas of charged particles, a plasma. This electrically charged gas will take the form of its container, yet the particles are not totally free to wander in the sense of neutral gas; instead, oppositely charged particles are attracted to each other. It's like a gas consisting of tiny magnets, forever attracting and repelling each other. This means plasmas have a higher density than a normal gas and in some sense maintain a loose shape or flow ('plasma' comes from the ancient Greek meaning a mouldable substance). The plasma state satisfies both the observed temperatures and the density of the Sun, as plasmas can still behave as gases at densities that would normally be high enough to indicate a liquid.

One of the key challenges for Earth-based nuclear fusion is reaching temperatures of 150 million celsius in order to heat the plasma enough to ignite fusion. We do not have the gravitational potential energy of the Sun at our disposal, and currently the methods used to heat the plasma utilise more energy than the energy produced by the fusion. Still, with ever-more efficient heating methods and the first commercial-scale fusion facility, the International Thermonuclear Experimental Reactor (ITER),[13] under construction in southern France, the future for fusion looks bright.

A star is born

We finally have a star, with its pressure and gravitational forces in harmony. To get to that point has been a battle. Many gas clouds never achieved the right temperature or mass to balance the gravitational and pressure forces. This balance dictates quite precisely what kind of cloud is the lucky kind, the kind that gets to shine for billions of years. The lower mass bound comes in at about 8 per cent the mass of the Sun and is so well defined because we know a lot about the energies required to ignite nuclear fusion. Gas clouds smaller than this mass limit may begin to collapse, but there simply isn't enough pressure at the centre of the star to fuse the hydrogen atoms together. As promising as it looked, this protostar is a failed star, or what we call a brown dwarf. Brown dwarfs are too large to be called planets and too small to be called stars, and recent research indicates that there may be as many as one brown dwarf for every two 'proper' stars in the Milky Way.[14] That's a lot of unlucky clouds. Initially glowing because of the energy release triggered by gravitational contraction, they are left to cool and fade, setting for all eternity.

The upper mass limit on stars is open to more debate, especially when it comes to the first stars. In theory, the larger the star, the more volatile the radiation pressure at the centre of a star is. The energies are so large that they are unstable and even a small energy change, for example from the accretion of a neighbouring gas cloud, can cause large changes to the equilibrium between the gravitational pressure and radiation pressure. Effectively, the larger the star the more likely it is to exhibit diva-like tendencies under pressure and throw tantrums, suddenly expanding and dissipating or suddenly collapsing. Because this behaviour is unpredictable, there is no fast rule for the maximum mass of a star, but when we study the stars around us, those above 50 solar masses are rare. When it comes to the first stars, though, there is reason to believe that this upper mass limit could have been a lot more relaxed than for current star formation. The pristine environment in which the first stars formed naturally lead to much larger stars, perhaps hundreds of times the mass of the Sun, as we will see in Chapter 6.

★   ★   ★

Stars are the result of a finely balanced dance of forces, and there are many failed clouds of gas that were unable to create exactly the right conditions for star formation. In the very first stages, the combined gravitational force of a cloud of atoms cause it to condense. If the cloud is not the right temperature or size, the kinetic energy of the atoms will easily overcome that initial gravitational push, dissipating the atoms. If the kinetic energy is low enough, however, the cloud will continue to condense, releasing radiation and thermal energy as the decreasing gravitational potential energy is conserved. If the cloud is massive enough, then the internal pressure will ignite nuclear fusion. Two hydrogen atoms fuse together to form a lower mass atom, releasing altogether a huge amount of energy. This energy is more than enough to explain the production of heat and light in stars, but the fuel cannot last forever.

The pressure of nuclear reactions holds up a star against its own inevitable demise, giving it a stay of execution. And it is a mighty long stay of execution. There's an awful lot of hydrogen to burn in a star and since the temperature is rising at the centre, more complicated nuclear reactions can be ignited one by one as the requisite energy is reached. The star steadily fuses its hydrogen, then its helium, holding itself up against collapse and producing the light and heat we see when we look at our Sun. For hundreds of millions or more often billions of years, a star can exist in equilibrium, with both teams of the tug-of-war in harmony and balance. But it cannot last, because gravity doesn't ever tire. Gravity remains, doing its work, trying to collapse our star even further. While there is a lot of hydrogen to fuel the counteracting nuclear fusion pressure, there is only a finite supply of it. One day it will run out and gravity will win. In our stellar game of tug-of-war there is only one dependable team in the end, as in the beginning.

CHAPTER FIVE

The Dark Ages

When I was working on my PhD, there was an artist in residence in our astrophysics group. She would sit at the back of our Monday seminars taking notes, while most of the audience fought the urge to fall asleep. Our lethargy was no comment on the quality of the speaker and instead reflected the tendency for researchers to stay within their niche. Work on planets in other stellar systems? Who cares about how the Universe began? Work on cosmology? Do we really need to hear about another alien planet? Katie Paterson seemed to care about it all, though. True to form, as a PhD student I felt that I didn't have time to pay attention to the arts. As a result, what Paterson did with her residency passed me by, to my detriment. Almost a decade later I saw an artwork depicting the evolving Universe, and after some digging I found out that the artist was Katie Paterson.[1] Big Universe. Small world. What Paterson did was to create a spinning wheel with continuous colouring depicting the overall colour of the Universe at different times in its history. You can look at a still image of the artwork, *The Cosmic Spectrum*,[2] in the colour centre-fold.

At first the Universe was hot, white hot with the energy of the Big Bang (just before 3 o'clock on the

colour image). As the Universe expanded, it cooled
down, through blue, yellow and orange and into the
longest wavelengths of visible light, the deepest, darkest
of reds. This was the Dark Ages. The Universe had
expanded and the gas had cooled down, so that it barely
emitted light at wavelengths within the optical portion
of the spectrum, if at all. It was before the stars had
formed, when the gas was still coalescing to ignite
fusion. Then, suddenly, a flash of blue as the first, massive,
hot, blue stars came into existence and heated the gas
with intensity in the Cosmic Dawn. They lived such
brief lives, however, and soon the second and third
generations of stars came into existence, cooler, yellower,
until we reached the colour of our Universe now.
Scientists have figured out the colour of our Universe
by averaging out the colours of observed galaxies,
adjusting for the redshifting effect and also for how our
eye is more sensitive to some colours than others. The
most likely colour a human would see if they mixed
all the colours together is beige, which they called
'cosmic latte',[3] presumably to make beige interesting in
a press release. It's a rather disappointing hue, however,
considering the rich depth of colour and variety out
there. The wheel goes further, exploring what our
Universe will be at the end, as the stars all complete
their lifetimes as red giants. And when the last star has
faded, and the Universe has expanded so much that
new stars can no longer form from the diffuse and
stretched gas: black for eternity. Bleak, eh? Let's turn
back the wheel and concentrate on beginnings instead
of endings: the Cosmic Dawn.

Taking the temperature of the Universe

Imagine that time when the darkness was absolute. True darkness. Darkness where your brain still forces you to look around, despite the futility. The chances are that unless you have sought the experience, true darkness has escaped you. In the modern world artificial light bombards us. Ninety-nine per cent of Europeans and North Americans live under light-polluted skies.[4] Darkness has become a rarity and even a commodity to sell. For the right price in cities around the world, you can relax in floatation tanks, eat your dinner or even speed date in complete darkness. Those of us wishing for less extreme versions of these experiences might join amateur astronomers and interested campers, and instead head to the hundred or so International Dark Sky Places. There the skies are kept free from the glare of lit offices and the haze of unshielded lamp-posts, enabling us to look at what has been above our heads the whole time: the planets that potter across the sky, the stars that configure to tell us stories and, if we are very lucky, the Galaxy splashed across the sky. This celestial mural reminds us that, try as we may to blot out the proof, we are very small. Looking into the darkness of the Dark Ages now serves to remind us we are also very young.

To calculate the colours of the early Universe, Paterson worked with scientists to work out the temperature of the Universe at a point and assumed the Universe was radiating as a blackbody, which we have seen is a reasonable assumption given the blackbody nature of the Cosmic Microwave Background, or CMB. The blackbody spectrum is characterised by temperature only – given the temperature you can know the intensity of light emitted over all wavelengths, and average them to get an overall colour.

Light gives us a serendipitous way to look back in time. Because of the finite speed of light, the further away the object we are looking at, the further back in time we are seeing it: the Sun appears as it was eight minutes ago, Mars appears as it was four minutes ago and the Andromeda galaxy appears as it was 250 million years ago. Ideally we would look far away enough to observe the first stars forming around 13 billion years ago by detecting their optical light. The technology to build an optical telescope able to operate with that sensitivity is beyond us, so we have to come up with another way, and the answer is in Katie Paterson's *Cosmic Spectrum*. What the wheel tells us is that we can track what is happening in terms of star formation by looking at the colour, or temperature, of the gas hanging about in the Universe. As those lucky clouds of gas are fusing, the unlucky ones are dissipating to make up what we call the interstellar medium: a diffuse collection of mostly hydrogen and helium atoms drifting about in the space between our first stars. If the gas is warm, star formation is likely to have established. If the gas is cold, then there aren't any stars about, because they haven't formed yet, or because they've all died out. What if we could find the first stars not by seeing them, but by measuring when they heat the gas enough to give away their positions?

Several experiments have been attempting to do this over the last decade. We call them global experiments in the sense that 'global' means including everything, or applicable all round, because they seek to constrain the

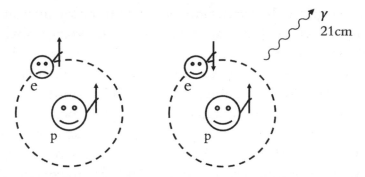

Figure 13 *The spin-flip transition. If an electron undergoes a spin-flip transition, its spin changes from parallel to antiparallel and the atom emits a photon (γ) of wavelength 21cm.*

era of the first stars by averaging the observed temperature of the sky. The Penzias and Wilson observation of the CMB was also a global measurement: they detected a single temperature as opposed to building a map of temperature variations across the sky as later missions like WMAP and Planck did. To make that measurement of the CMB, the scientific descendants of Penzias and Wilson tuned in their telescope to the frequency where they knew characterised the CMB. We apply the same principle when detecting the temperature of the primordial gas. Hot gas emits radiation – in particular hydrogen gas emits radiation at 21cm in wavelength.

The spin temperature
We call it 21cm radiation because the photon has a wavelength of 21cm, or a frequency of 1,420MHz. A hydrogen atom emits a 21cm photon when its electron changes spin state. Spin is a characteristic of atoms, in the same way that mass and electric charge is, but it is less familiar to us as it has no analogue on our scale. When

the proton and electron have their spins parallel, they are in a more excited state; the system has more energy. If the electrons and proton spins are antiparallel, the hydrogen atom is in a less excited state. If the electron changes spin from parallel to antiparallel, as it can do spontaneously or through external prompting, it sends out a parcel of energy in the form of a 21cm photon. The atom is then in the ground state. This is like having two N–S magnets side by side. It takes more energy to keep them touching N–N and S–S – they naturally want to be antiparallel N–S S–N. A hydrogen atom wants to be in its least energetic state, so if an electron has a parallel spin, it won't be long before it de-excites in a *spin-flip transition* and emits that 21cm photon. The *spin temperature* of the gas tells us what temperature the gas would need to be to have a certain proportion of its atoms in that excited state and the rest in the lower energy state. A higher spin temperature indicates that a higher proportion of the gas is excited into the higher-energy state – hence the greater the intensity of 21cm radiation that can be expected to radiate from the gas.

The spin temperature of hydrogen gas has an interesting history through cosmic time, going up and down according to what was affecting it most. In the early Universe, the hydrogen atoms underwent a spin exchange when colliding with hydrogen atoms or electrons with opposite spin, setting the spin temperature. As the Universe expanded, these collisions occurred few and far between, and instead absorbing CMB photons became the dominant force behind determining how many hydrogen atoms were in the excited state, thus setting the spin temperature. As the first stars formed, a third mechanism for setting the spin temperature came into play. The first stars emitted high-energy UV photons that both heated the gas and mixed up the spin temperature. Hydrogen atoms absorbed the UV photons, exciting the electrons above the energy level indicated by the spin-flip transition. Electrons do not favour high-energy states and as they come down to a lower energy state, the atom emits photons. Some electrons in some hydrogen atoms will find themselves falling into the excited spin-flip state, then releasing a 21cm photon in the last step down to the ground state. This mixing by UV photons is called the Wouthuysen-Field effect. It decouples the spin temperature from the CMB and instead couples it to the gas temperature. The UV photons are both heating the gas and setting the spin temperature, coupling the two.

The first detection of first light
Global 21cm experiments aim to measure the intensity of this 21cm radiation at different times in the Universe's history, watching it dive and rise as the Universe expanded and created the first stars. It's both a simple and difficult experiment: simple in concept, demanding in execution. I've given countless presentations on the front-runner telescopes for achieving the first detection of the first stars, but a global experiment had never featured in my slides. And yet it was that in February 2018 I received a phone call from a journalist,[5] requiring a comment on the first detection of the first stars by the Experiment to Detect the Global EoR Signature (EDGES).[6] EoR stands for Epoch of Reionisation, the period when the first stars heated and ionised the surrounding gas.

The paper was embargoed, which meant that I would have the privilege of reading it before most of my colleagues, and my stunned silence gave way to excitement. It was almost euphoria, and I jumped about in my house. Yes, it was a competing experiment to the one that I have worked with for the last decade. The excitement of finally achieving some data from this era, cracking open the door a little, felt shared, though. EDGES had taken the temperature of the gas 180 million years after the Big Bang. It had found that point on the Cosmic Spectrum wheel when the dark red of colder gas began to give way to the bright blue of a heated Universe, showing that the first stars had formed. They had found the end of the Dark Ages, the first rays of the Cosmic Dawn.

EDGES is a TARDIS of an experiment.[*] There are two experiments, scaled replicas, one tuned to the higher frequencies of the Epoch of Reionisation and the other, the low-band antenna, tuned to the lower frequency photons of the Dark Ages. Each comprises what is fundamentally a metal table. Two horizontal sheet antennas sit upon a support structure together with a metal groundsheet aimed to minimise interference from the already pin-drop-quiet environment of the Western Australian desert. I am floored every time I look at an image of one. A table! In the desert! And it can look back all the way to the beginnings of our Universe!

---

[*] I'm meant to explain every acronym I use, so: Time and Relative Dimension in Space. I can forgive you not knowing that but if you don't know where TARDIS is from, go watch *Doctor Who* and rejoice.

When you compare it to the fancy space-bound Hubble, it is humble in size, price and glamour – but not results. EDGES measured the temperature of the hydrogen over a period of about 100 million years, relative to the temperature of the background (usually assumed to be just the CMB). If the spin temperature exceeds that of the CMB, then we observe the temperature to be positive, or in 'emission'. If, however, the spin temperature is colder than the CMB, we observe the temperature to be negative, or in 'absorption'. At times in the timeline where the spin temperature is equal to that of the CMB, we cannot see it at all... the overall temperature measured is zero.

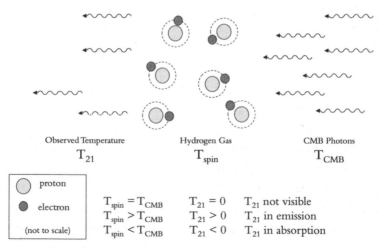

Figure 14 The 21cm radiation temperature. We observe the temperature of the hydrogen gas relative to the CMB. In the early Universe the temperature is observed in absorption but, once the first stars have begun to heat the gas above the CMB temperature, the signal moves into emission.

As the first stars were forming, the gas was at its coldest. The Universe was expanding, allowing the gas to fill a larger space and cool down. The gas in between the stars, the remnants of our unlucky clouds of gas, would produce a signal below that of the CMB, in absorption. EDGES took this temperature measurement across a range of observation frequencies. As these photons of 21cm wavelength, or frequency 1,420MHz, travel towards us they have to work against the expanding Universe: they are walking up an escalator the wrong way. This costs them energy, which is equivalent to lengthening their wavelength. We call this stretching of wavelength *redshift*, an effect we first saw in receding galaxies. We can put this redshift to use by tuning our telescope into the different wavelengths, or frequencies, of the stretched photons to investigate different times. A photon emitted 180 million years after the Big Bang will arrive to us now with a wavelength of 4.5m, or a frequency of 68MHz. A photon emitted 100 million years later will have been fighting against the expansion for less time, so it has lost less energy and arrives to us with a wavelength of 3.3m, or a frequency of 92MHz. EDGES detected radiation around these frequencies, and it measured a trough in absorption. What this trough shows is that before the first stars were forming, point (a) in Figure 15, the spin temperature was at the same temperature as the CMB, so the measured temperature was zero. At 180 million years after the Big Bang, point (b), EDGES detected a sharp absorption: the spin temperature had decoupled from the CMB. Instead, it coupled with the gas temperature because of the UV photons of the first stars both heating the gas and mixing the spin states through the Wouthuysen-Field effect. This sharp absorption was the signal we had been looking for – the point at which the UV photons became the dominant mechanism for setting the temperature of the gas. The first stars had come to life 180 million years after the Big Bang, and would soon heat the gas so much that it would match, then exceed, the temperature of the CMB, point (c). We could now add a label to the end of the Dark Ages in Paterson's spectrum with a lot more confidence. This is the first direct signature of the first stars we have. A sharp dip opens the curtains on a hidden age and gives us the most basic morsel of information: a date of birth.

The unexplained signal

If the story of EDGES ended there, it would have still made the news and I still would have jumped around the house but it wouldn't have been potentially revolutionary. As I sat down and read the rest of the paper, my excitement changed to a quiet confusion. EDGES had found something odd.

When the signal goes into absorption (when the spin temperature is less than the temperature of the CMB), it should be a quick dip down only, as the same UV photons that couple the spin temperature to the gas and drag it down also heat the gas and drive both of them upwards. Every known model of the global 21cm signal has failed to reproduce the shape and magnitude of the detected EDGES signal. This is the most exciting and painful thing that can happen in science, though it depends on how close to the data you are. For those of us looking on, there is the excitement of discovering

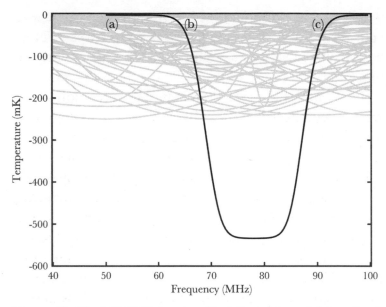

Figure 15 *The EDGES detection (black line). No model (grey lines) constructed before February 2018 could reproduce the depth of the trough, and none to date has been able to replicate the flat bottom.*

something unforeseen. For those responsible for the data there is an uncomfortable burden: the extraordinary proof required to prove the phenomenal. When you read the EDGES experimental paper,[7] this burden is clear. There is hesitancy throughout and the principal message is clear: 'we still seek confirmation'. The result is described clinically, mirroring the treatment that Penzias and Wilson gave their CMB paper 53 years before. The authors don't explain what could cause this unexpected signal and instead offer a comprehensive list of things it cannot be. They followed the signal over two years of data, detecting it over different frequencies, with two different instruments and with antennas in different alignments. They changed some hardware to be sure it wasn't a source of noise, used different analysis pipelines and observed at different times of day and year. They presumably checked for pigeons. And still, the signal remained.

Those involved in EDGES kept their detection secret while they checked and rechecked the technology and science behind it. In Croatia in 2017, astrophysicists met to discuss progress on the detection of the first stars. I sat and listened to updates from all the major telescopes, and I remember listening to cosmologist Judd Bowman's presentation on EDGES. Or rather the lack of it. At the time I made a note that he had instead concentrated on the next-generation telescopes, with very little mention of EDGES at all. I regarded it as a dismissal of a stalled experiment as opposed to the silence of a successful one undergoing confirmation. In retrospect, I can understand the silence. They had to be sure because the implications were so extraordinary.

The EDGES signal was more than twice as cold as anyone had predicted. Mirroring the Penzias and Wilson CMB announcement, a theory paper accompanied the release of the EDGES experimental paper. In the theory paper,[8] experimental hesitation gives way to the theoretician's enthusiasm. The lengthy list of what this signal is not gives way to an unorthodox idea of what it could be. To make a gas colder, it must interact with something colder, and 'the only known cosmic constituent that can be colder than the early cosmic gas is the dark matter'.

Dark matter

Did I forget to mention dark matter before now? Oops. It's quite an omission to make, but it reflects the state of the field of the first stars before 2015, as it reflects the state of the lecture hall I sat in alongside Katie Paterson, as a PhD student in 2011. Dark matter? Who needs to know about that for the first stars? I must ask Katie for her notes.

In previous chapters I spoke of a Universe in which hydrogen was the main constituent, but my wording was careful. I described the normal Universe, the Universe that we could 'see', even if our seeing required an X-ray or radio telescope. 'Normal' matter is made of baryons, which interact with light so that we can see their presence, using some part of the electromagnetic spectrum at least. Dark matter doesn't interact with light, so escapes all of our traditional telescopes and even our most modern ones. We have never directly detected dark matter and this is no insignificant thing as we think dark matter makes up 85 per cent of the total mass within the Universe.

It's not that first stars scientists disregarded dark matter entirely – more that we thought that dark matter could only noticeably interact through gravity. Dark matter creates a factory floor on which to construct the first stars. In the early Universe, after recombination 380,000 years after the Big Bang, the gas was hot, too hot to collapse into anything approaching a star: there were no lucky clouds of gas at all. However, to the best of our knowledge, and there is some argument about this, dark matter was cold. Cold stuff can collapse, so collapse it did. Our models show that the dark matter condensed into dense filaments, a cosmic web covering the entire Universe. It didn't collapse further, say into stars or galaxies made of dark matter because, while it was cold enough to condense a bit, it cannot interact electromagnetically (that is, with photons) to lose heat like baryons can. Because dark matter couldn't lose this heat, the internal pressure was much higher. This kept the dark matter from condensing into anything much smaller than regions of gravitationally bound dark matter, called halos, of about 1 million solar masses, at the junctions of the filaments.

Once the cosmic dark matter web has formed, the cooling gas has a substrate to cling on to, gravitationally. The gravitational pull of these clusters of dark matter causes that gas to condense along the same pattern. One of the best simulations of this cosmic web formation is the Illustris simulation,[9] a slice of which can be seen in the colour centre-fold. This is an image of the most massive cluster of dark matter in the simulation, at a filament junction. It contains more than 5,000 sub-halos: halos that can form galaxies within. On the left of the image is the dark matter distribution; on the right is the gas distribution. The two blend seamlessly. This is one of my favourite astrophysics images. It is my desktop wallpaper. In place of family photos I have an ultrasound of the baby stars, because that is what this is. The dark matter brings the gas together, pulling it in to form clouds of gas that birth stars. The gas follows the dark matter perfectly on the large scales, but on the smaller scales, zoomed right in, the gas can keep collapsing where the dark matter cannot. While the dark matter stopped collapsing because of a lack of a cooling mechanism, the gas can continue its collapse in the way described in Chapter 4. Thus the first stars and first galaxies formed in these dark matter *halos*, a bright light in the centre of a cloud of dark matter.

Flat rotation curves

Our own Milky Way sits in the middle of a dark matter halo, as almost all galaxies do. In fact, it is that underlying halo structure that led to some of the most convincing evidence for dark matter. In the 1970s scientists could look at a galaxy with enough resolution to measure the velocities of stars using the Doppler effect, all the way out to the edges of that galaxy. The Doppler effect is the shifting of emitted light to a longer or shorter wavelength, according to whether the source is moving away from us or towards us, as shown in Figure 10, Chapter 3. American astronomer Vera Rubin plotted how the velocities of the stars changed according to their distance from the centre of the galaxy. In a galaxy made only of normal matter, we would expect a *Keplerian* descent, whereby the velocity of the stars falls with distance. Further out from the centre of the galaxy the gravitational pull of the luminous mass attenuates, and the orbital velocity of the stars should therefore lessen as a result. This is what we observe in the Solar System. Eight planets are in orbit, but as you go out from the centre the planets not only have a longer path to travel but also move more slowly around their orbits. Earth orbits the Sun every 365 days. On Mercury one orbit takes 88 days, while on Neptune it would take 165 years.

Instead of the expected Keplerian descent, Rubin measured flat rotation curves for almost every galaxy she observed. At first, the velocity increased with radius of measurement. This was in line with expectation as more of the main bulk of luminous mass was encapsulated within the increasing radius and so the gravitational pull increased. However, once observations breached the

outer regions of the luminous galaxies, the curves quickly
flattened. And vitally, instead of then beginning a descent,
they remained level. Flat rotation curves, in almost every
galaxy she observed. The flat curves imply that as you
observe further from the centre of the galaxy, the stars
still move just as fast. The only way this can happen is if
the amount of mass is increasing as you go further out,
so that the gravitational pull on the stars is maintained.
This doesn't align with our optical view of a galaxy:
galaxies have bright centres and diffuse outer halos, not
the other way around. The conclusion can only be that
there is a colossal amount of matter in the galaxy that we
can't see: dark matter.

Dark matter is everywhere. We are sitting in a great
big ball of it right now. We don't know about it because
it doesn't seem to interact with anything apart from the
gravitational force, and its density is so low that on human
scales we can't feel its pull at all. Over our lifetimes, only
about 1mg of dark matter will pass through our bodies.[10]
This low density is why we have a Keplerian descent in
the Solar System and not a flat rotation curve. The dark
matter is there, but its density is so low that it makes little
difference. While Neptune feels the gravitational force
exerted by all the inner planets *and* all the enclosed dark
matter, the latter only makes up the mass of a large rock.
On a Galactic scale, though, all that matter adds up.
In the outer regions, as you increase the radius you
increase the volume massively, encapsulating only a few
more visible stars but a huge amount more of dark matter.
At that scale we finally see it at work, pulling on those
outer stars and making them go faster than we thought
possible.

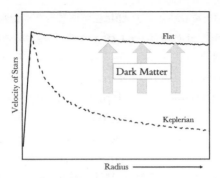

Figure 16 *Flat rotation curves. The observations made by Vera Rubin and Henry Ford in 1978 found that the galaxies they observed followed a flat rotation curve, where the rotational velocity of stars remained the same despite the increasing distance from the centre of the galaxy.*

The dark matter particle

But what is the stuff? We know there is something creating a gravitational pull and that its density is low. One of the more mundane theories begs us not to get too carried away by the exotic. Just because we can't see it, it doesn't mean it's this strange new mysterious stuff. It could simply be … dark. We talked before about the number of failed clouds of gas and in particular the multitudinous brown dwarfs, failed stars. Maybe when we measure the mass of a galaxy, it is also full of brown dwarfs that add a lot of mass too. It makes sense, but it isn't a convincing theory. Paterson drew on dead stars lying unseen in our Universe in her work *All the Dead Stars*.[11] She placed a point on the map of the Galaxy for the 27,000 known dead stars ever recorded in human history. Despite stars being dead, faint or failed, we find a lot of them now. We have very good telescopes, and even though brown dwarfs are faint we can detect enough of them in our galaxy to calculate the masses involved. It doesn't dent the amount of missing mass purported to be dark matter.

One theory is that dark matter is a new member of the known particle zoo of protons, neutrons, electrons, quarks, neutrinos and more. We know it doesn't interact in any large way because we would have noticed, so the hypothetical particle is named the *weakly* interacting massive particle (WIMP). WIMPs represent the hope that just because dark matter doesn't interact with other particles a lot, it doesn't mean that it doesn't interact at all. One dark matter paper constrained these interactions based on exploding humans, in a paper entitled: 'Death and serious injury from dark matter'.[12] The authors proposed that if dark matter is made of particles each with a large mass, then collisions of a particle with the human body would cause ... a mess. That humans don't explode in unexplained circumstances on a regular basis tells us that the dark matter particle must be below a certain mass and have a low probability of collision. If it were more massive or more strongly interacting, the collisions would be prevalent and more destructive, and we would have noted the phenomenon by now. Wonderful. One of the earliest and continuing authorities on first stars science is Professor Abraham Loeb. Loeb is the chair of the Astronomy department at Harvard University and a highly respected and enthusiastic scientist. When we spoke in 2019 he lamented the tendency of scientists to lose the curiosity we have as children. 'We should behave more like kids. I was trying to encourage my colleagues here at Harvard to behave more like kids ... Mistakes are part of the learning process.' I loved that take on science – it should be fun and curious, and we should follow the questions and be unafraid of there being dead ends or surprise results. The 'Death by dark matter' paper is such a paper, and EDGES was a path that few went down but that ended in a tremendous surprise.

Several experiments around the world are attempting to amplify the tiniest of dark matter interactions, if one ever happens, to a level that we can record it, achieving the first direct detection of a WIMP. These experiments lie deep underground all over the world: Britain, Canada, the United States, Australia, Spain and Italy. Sheltered from the particles swarming above the ground, the experiments house materials whose nuclei are predicted to recoil when a WIMP collides with them. The slight recoil produces light and heat in the most minuscule of parcels, but with sensitive detectors, cryogenic temperatures and a lot of patience, scientists believe that one day, if a dark matter particle has a rare collision, they will detect it. So far, they haven't. These experiments have been running for more than two decades now and there has only been one, highly contested, positive result.[13] The majority have detected nothing at all, continuing the dearth of data on dark matter. But then, where we last expected to find information on dark matter, there was EDGES.

Dark matter and first light
Now we have a bit of background on dark matter, we can get on to how it could be connected to the EDGES observations. Before we do though, I would be remiss not to present a plea for caution. If you study the literature, there are a few concepts that could explain

the EDGES signal, the most sensational of which is interacting dark matter, because of the unexpected and far-reaching consequences for another scientific field desperate for data. So while we should allow ourselves to feel the excitement of the possibility, the following should be read with the healthy skepticism of a scientist. It is one of a few ideas, and we need more data before we can present it as a robust theory. We have known about the connection between dark matter and gas and galaxies and stars for a long time, and we have included it in our simulations, so it's not as if we ignore dark matter completely. Once the gas has coalesced into a star within the web of dark matter, however, we have little need to consider dark matter. It's on the wrong scale to matter, just as dark matter has little effect within the scale of our Solar System. Or so we thought. What the EDGES result implies is that something is interacting with the gas to make it colder, and the only thing hanging around in the Universe colder than the gas is dark matter.

Given a hot cloud of gas and a cold cloud of dark matter, energy is passed from the former to the latter until the two are in thermal equilibrium: the gas cools and the dark matter warms up a bit. But hang on – dark matter is famous for not interacting with stuff; it's even in the name WIMP. What the EDGES theory paper pointed out was that we are in a very different Universe now compared to way back when. What the 3pm blue of Paterson's wheel shows is that, several hundred millions of years after the Big Bang, the gas in the Universe was at the coldest it had ever been and would be still continuing to the present day. Soon the first stars and all the subsequent generations of stars would heat the gas and keep it cosy. Nowadays dark matter doesn't interact much or even at all with the particles around us, otherwise we would have seen the evidence by now, either in a deep underground experiment or in the frequency of exploding humans. But what if dark matter only interacts at very low velocities, or low temperatures? There are lots of different particle interactions or collisions, some of which will only happen in high-energy collisions, while others require a more calm environment. Try hugging someone sprinting in the opposite direction and you'll understand why sometimes slower is better when it comes to interactions. The Dark Ages provide the calmest of environments, when the gas particles were moving at their slowest speeds. Therefore if dark matter collisions are most likely in cold environments, this is when they would most likely happen. This neatly explains why dark matter is so unsocial now, yet was so much more interactive in the past.

There are other constraints on dark matter, from the CMB, for example, so any theory has to satisfy those constraints while creating the cooling effect in the EDGES data. Satisfying both has turned out to be hard, and only a small range of cold dark matter models are viable. One that stands out is the millicharged dark matter model.[14] If dark matter has a small electrical charge, it could interact with the protons and electrons in the gas to cool it down. This sounds great, but when something is charged, magnetic fields affect it as well as gravitational ones. We can constrain the models by considering what would happen to this kind of dark matter in our Universe and compare with our observations. The magnetic fields associated with supernovae (stellar explosions) would repel and expel any millicharged dark matter from the galactic disk and in addition, the galactic magnetic field would prevent that millicharged dark matter from reentering the disk of the Galaxy. Our measurements of the mass of dark matter in the galactic disk indicate that most of the dark matter, in fact more than 99 per cent of it, has to be uncharged, because it evidently stuck around in the face of the magnetic fields produced by supernovae. Had it all been expelled, we wouldn't see those stars moving as fast as they do in the outer regions of the Galaxy. Perhaps our understanding of dark matter doesn't have to change too much; perhaps it is only less than a percentage that is more mysterious and maybe more charged than we thought. Perhaps dark matter underwent its own version of recombination (the event in the early Universe when protons and electrons combined to make atoms), forming neutral dark atoms (the 99 per cent of dark matter we observe now), but leaving a residual number of 'free' charged dark matter particle constituents (the millicharged dark matter). This is a field that is moving so fast that these paragraphs will need revisiting over the next few years. Whatever EDGES tells us about dark matter, it is the fact it tells us anything at all that is most surprising. EDGES was going to tell us about the first stars, and it has, but we've almost lost that success in the unexpected bonus of potentially exploring dark matter.

The trouble with troughs
When the EDGES results were released the newspapers lead with the dark matter explanation, understandably given the focus of the sole theoretical paper that

accompanied the experimental paper. Over the months following, however, other papers appeared with speculations. An excess background or a mistake in the signal calibration, for example. To understand the former requires a change of perspective. What if instead of the gas being colder than expected, the background was warmer: an extra background on top of the CMB. For the maths to work out, any additional background would need to be equal to or greater than the background provided by the CMB, a significant change. One of the first papers to propose the excess background theory is infused with the same excitement as the first dark matter theory paper.[15] There is even an exclamation point in the text, which I'm not sure I've ever seen in a scientific paper before. It's simply not done. I love seeing this enthusiasm and humanity in scientific papers, but there are others who would argue against me. They would say that science should be dehumanised, clinical and purely logical. I disagree. When you read the older scientific papers from the 1700s to the twentieth century, this excitement is present, alongside admissions of mistakes or confessions to not knowing where to go next. In the twentieth century, it's hard to pin down when we lost this. This lack of admission of humanity has led to a constant inflation of the success of results, a grandiosity that leads to the smallest of results lauded as groundbreaking. It detracts from collaborative thought and reduces science to branding. I don't like it. I love it when I access these new EDGES papers, and feel my colleagues' excitement and read that they've tried really hard, but they still aren't sure. They must use 'optimistic assumptions' to get a theory to fit, and it forces them to make an 'extraordinary claim that would require extraordinary evidence'. There are 'significant issues' new theories 'must overcome'. Everyone must bend their long-held assumptions and make room for a new collection of theories.

The excess background has precedent. Two experiments, the Absolute Radiometer for Cosmology, Astrophysics and Diffuse Emission 2 (ARCADE 2)[16] and the Long Wavelength Array (LWA),[17] have detected unexplained backgrounds, albeit at different frequencies. If these backgrounds are real and were present at the frequencies of the Cosmic Dawn, they could explain the trough. One theory with beautiful symmetry is that the signature of the first deaths impacts the signature of the first stars. Population III stars are thought to end their lives as black holes or in gigantic explosions called supernovae. Both produce high-speed particles that produce radiation, called synchrotron radiation, as they accelerate. This creates another background of photons, but it also produces the opposite effect to what we want. Black holes produce X-rays that heat the gas, lessening the absorption. Therefore for these whispers from dead stars to be the background, we also have to have an environment where dense gas surrounds the stars. In this way, the X-rays from the supernovae and black holes cannot get out to heat the wider gas. There are ways to recreate the signal using an excess background but, as with the dark matter theories, a lot of fine-tuning is required to get there. The field of study concerning black holes is one that always needs more data, so there is excitement that the EDGES trough could indicate the early presence of black holes. At the centre of our galaxy, and most spiral galaxies, lies a supermassive black hole, with a mass about equal to 5 million Suns. Such black holes are so large than we cannot figure out how they have grown so big in 'only' 13 billion years. If Population III stars formed black holes much earlier than we thought, as early as the Dark Ages, then that could give the black holes enough time to grow to the proportions we observe today. Another cosmic mystery potentially solved by EDGES.

★ ★ ★

We've heard the case for dark matter and for an extra background. And so we come to the third possibility that the detection is inaccurate. Despite the simplicity of the EDGES design, the data analysis is complex. When you switch on a radio telescope and tune it to a frequency, you don't only pick up the photons of that frequency from the first stars. You pick up *all* photons of that frequency, including Earth-produced ones from mobile phones, planes and radio stations. The Galaxy itself is an excellent source of photons and they outnumber the first stars' signal by hundreds to thousands of times over. Establishing how to get the signal out from under all those foregrounds is my career speciality. It requires statistical methods and excellent knowledge of how bright the Galaxy is at each frequency. We're good at figuring out these things, but there is always the worry that there is an unknown unknown, perhaps a source of photons that contributes a foreground we aren't expecting and don't even know is there. A detection could be lingering foregrounds masquerading as the real deal. This theory has received increasing attention in the literature. At a conference after the EDGES announcement, I went up to one of the scientists involved in the EDGES collaboration and said I wanted to talk about foregrounds. He looked at me with an exhausted look and sighed 'Of course you do.' I think many people had been asking him the same questions at that meeting. Had they checked their algorithms? Had they tried a different method of foreground removal? The answer was invariably yes, as the thoroughness of the original paper suggests. It is not a self-satisfied announcement of what the detection is, but a self-conscious list of what it is not. They have tried everything they can to explain this signal with a mundane solution. We are at a stalemate of theories that can only be resolved with more data.

The end of the Dark Ages?

We find ourselves in the Dark Ages in more ways than one. Historically, the Dark Ages referred to the period approximately between 500 and 1400 AD. It is a term that conjures up barbarians, the Black Death, and a lack of education, comfort and culture. Italian scholar Petrarch coined the phrase in the fourteenth century, using it to refer to the lack of literature in the previous centuries. Despite being based on one man's annoyance at not being able to find a good book, the term stuck. As time has gone on, plenty of evidence has been uncovered of scientific progress, arts and cultural enjoyment during the Dark Ages. As a result, the term has fallen out of favour, apart from by a few historians who keep it to refer to the lack of records during that time. They have extended this latter interpretation to a more modern age: the digital dark ages. This is the perceived dark ages cause by an inability to process or parse outdated file formats where no hard copy remains. In the digital age, data is so often deleted or recorded in formats we do not understand how to access (ask a teenager to load the data from a floppy disk). In the sense of missing data, we are leaving the dark ages of the Dark Ages. We have a little data and we have found a single historical record that has already caused us to rewrite the history we thought we knew. Akin to the discovery of a floppy disk in a medieval archaeological site, EDGES has made us question a lot that we held to be certain.

There are other telescopes that have shifted their focus to the frequencies of the absorption trough, to get more data to verify and validate the EDGES signal. These telescopes in separate hemispheres use different algorithms to analyse their data. Therefore the hope is that, if the detection results from a bad algorithm, it won't be repeated in the data from another telescope, so we'll know there is a problem.

The last of Paterson's art that I want to introduce to you is *Earth-Moon-Earth*.[18] For this work, Paterson converted Beethoven's *Moonlight Sonata* into Morse code. She bounced the signal off the Moon, where the surface itself and the atmosphere of the Earth corrupted the signal. Converted back into music, and played by a ghostly automatic grand piano, the Moon's version of the sonata is recognisable but dissonant, an uncomfortable listen. The clarity of the signal was compromised a lot by that 2.5-second journey, so you can imagine what a signal that has travelled 13 billion years might sound like.

We have the modern Universe's version of the Cosmic Dawn sonata, but it's our job now to figure out the original composition.

★ ★ ★

EDGES seems to have cracked open the door on to a previously hidden era. It has used the temperature of the hydrogen gas in the early Universe to trace the underlying presence of Population III stars. As fusion in the first stars began in earnest, the surrounding gas was heated, raising its temperature in comparison to the background CMB. The EDGES data suggests that the first stars had formed by about 180 million years after the Big Bang, but it also suggests that the hydrogen gas was much colder than expected. This could be due to dark matter unusually interacting with the hydrogen gas, but it could also be due to an excess background on top of the CMB. The experiment is complex and we need more data to validate and verify the detection before we make any strong conclusions.

A lot more work is needed to clarify a heavily corrupted signal and find out which theory is correct. The theories are not all exclusive. Perhaps a bit of all of them is needed to uncover the true Cosmic Dawn signal? And then perhaps something new no one has even thought of before? After all, to date, no theory can explain the flattened shape of the trough. Perhaps we will need to advance our theories into a new key altogether. We do not understand. Isn't that exciting?

CHAPTER SIX

Fragmenting Stars

There is a picturesque region called the Lake District in the United Kingdom. It is England's largest national park, a UNESCO World Heritage Site that is home to numerous mountains and, of course, lakes – 16 of them in all. The lakes are large enough to enable you to enjoy watersports and spend pleasant days sailing on them, assuming you get a day with pleasant weather (never a given in the UK, let alone in the Lake District). I had a good time there but couldn't help feeling a little underwhelmed by the lakes themselves. The reason? I have been to the United States. The North American Great Lakes contain just under a fifth of the Earth's fresh surface water.[1] One of the largest lakes in the Lake District, Windermere, has a surface area of near enough 15km² (5.79mi²), while Lake Erie, only the fourth largest lake in the US by surface area, covers 26,000km² (10,038mi²), or more than 1,700 Windermeres. To misquote Crocodile Dundee: 'Call that a lake? THIS is a lake.'

Lake Erie was one of the lakes I skipped on my tour of the US, and I'm not regretful of the omission, having read some health warnings accompanying the lake in recent years. In 2014, half a million people in the surrounding area were warned to avoid drinking tap water due to a toxin produced by algae that had taken over the lake,[2] microscopic blue-green bacteria called *cyanobacteria*.

You might not be on a first name basis with cyanobacteria, but you'll have come across them at some point, most likely when strolling around a body of still water. Despite their name they can be a range of different colours. When allowed to grow to excess they 'bloom', most often producing a green (though sometimes red, yellow, brown or blue) covering on the surface of the water. They don't look too pretty, they're not pleasant to swim through and they can be poisonous. They create a carpet that sunlight cannot easily penetrate, starving the life-forms below of heat and light, and depleting the oxygen levels. In addition, they can produce toxins that are fatal to some life-forms and would cause serious illness in humans if consumed in a glass of water. Despite how irritating we find them now, we have a lot to thank them for. In fact, life on Earth as we know it wouldn't exist if it wasn't for their intervention.

The great oxygenation event
More than 2.5 billion years ago, the Earth was toasty warm and replete with greenhouse gases such as carbon dioxide and methane. These gases trapped the incoming sunlight, warming the Earth. The oceans were at temperatures of around 65–80 °C, or 150–175 °F, which you would find more than a little uncomfortable if you took a paddle. For reference, the temperature of your bath, and your body, is generally around 37 °C, or 99 °F. There was no oxygen in the atmosphere to breathe, so the complex life that we see today had not yet evolved and could not have survived. Then, something

happened: the Great Oxygenation Event, or the Great Oxygenation Crisis, depending on whether you are cyanobacteria (the former), or everything else (the latter). It is proposed that cyanobacteria are both the instigators and the few survivors of one of the biggest mass extinctions in the Earth's history.[3] While cyanobacteria had been around for a little while, we think they evolved the ability to photosynthesise at about this time[4]. Photosynthesis is a chemical reaction between water, carbon dioxide and sunlight that produces energy for a plant in the form of glucose, as well as a waste product of oxygen. As all the cyanobacteria began to pump out waste oxygen, it was at first sunk back into other chemical reactions. It was used in the decomposition of dead organisms and it rusted the dissolved iron in the world's oceans, leaving layers of red sediment that are known as red beds or ironstones. Once we had a rusty Earth, the oxygen had nowhere else to go but up into the atmosphere. There, it reacted with the methane to form carbon dioxide and water, bringing down the methane levels as the oxygen levels continued to rise and rise. Cyanobacteria changed the atmosphere and thus changed the world. Once oxygen was in abundance, a new form of life could evolve, one that used oxygen as a fuel and would one day produce the gigantic, complex gatherings of cells we call human bodies.

In the early years of the Earth's history, the release of large amounts of oxygen into the atmosphere permanently changed the environment and what could thrive in it. The first stars played a similar role for the early Universe. They found themselves born into a pristine environment devoid of larger structure. The complexity we see around us today requires bigger building blocks than just hydrogen and helium, and back then it was only the first stars that could provide them. If we compare the scale of a meagre cyanobacterium against that of the Earth, it isn't all that different from the comparison between the first star and the size of the Universe. To the Universe, the first stars might as well have been bacteria: insignificant and immaterial.

The Jeans mass
In Chapter 4 we left our lucky cloud of gas at the point of fusion. The cloud had collapsed under immense gravitational pressure, pushing the very atoms of hydrogen together until fusion, hydrogen burning, was initiated, producing heat and light, and a pressure that pushed outwards and balanced the inexorable gravitational push. In Chapter 4 we zoomed in on the production of one star. One star from one gas cloud. It's a neat, symmetrical idea. Until a few years ago that's how we were simulating these first metal-free protostar formations: zoomed in, focused on one main star at the centre of the cloud. We didn't have the computational power to do more than that. The power it takes to compute the gravitational forces, the chemical inter-actions and the magnetic fields of just one star is at the very limit of computational capacity. As a result, a few years ago this chapter would have been called 'Lonely Lives', the same title I used to give to my slides on the subject. We thought that the first stars formed singularly and lived isolated lives. But we now believe that they were not lonely at all, and quite to the contrary formed in families, just like the stellar nurseries we observe today.

When we use infrared light to peer through the visibly opaque gas from which the stars form, we see a single cloud of gas has not formed a single massive star, but instead it has fragmented and collapsed into a cluster of smaller mass stars.

What, then, determines how many stars a cloud can form? The mass that a system must reach to collapse is defined through a quantity called the Jeans mass, named after British astronomer Sir James Hopwood Jeans. Jeans contributed to two major ideas that have been discussed in this book already. Once was in his contribution to our understanding of the blackbody spectrum, and the other was in his role as the first person to propose the steady state theorem of the Universe.[4] Jeans realised that the mass at which a system would become gravitationally bound would depend on how big and how hot the system was. A colder cloud will collapse more readily than a hot cloud because the atoms are less energetic and can less easily zip away. A denser cloud will more readily collapse than a more diffuse cloud because it squeezes the atoms into a closer space. The closer two masses are, the stronger the gravitational force between them, so the more force zippy atoms have pulling them back. The Jeans mass tells us that a star the mass of our Sun is likely to have condensed from a clump of gas about 250 times the size of the Solar System, assuming an average gas temperature of 20K.[5] In a hot system the gas atoms are whizzing about everywhere with high kinetic energy, so they need a larger gravitational force to slow them down and bring them inwards ... the Jeans mass will be high. If our Solar System had formed much closer to another star, then the gas would have been heated to 100K. Then, the radius of the gas clump would have had to have been closer to 1,250 times the size of the Solar System in order for it to collapse.

Once a cloud reaches the Jeans mass, it will contract. The clouds are heated so much by the contraction itself that they cannot break up and contract into little star-sized clumps: the Jeans mass is still too high, especially in a hot, young Universe. Only really massive clouds were large enough to overcome the thermal pressure and collapse. To break up that cloud we have to lose some of that heat and lower the Jeans mass in order that the cloud can begin to fragment into collapsing, stellar-sized clumps.

It seems odd to be considering how to cool down a gas cloud, to make a very hot star. The key is in the order of events. Gas clouds start out as very diffuse entities, and fragile ones, too. A star is a much denser object, so we have to work out how to condense a gas cloud that much, without it blowing itself apart by the increasing thermal energy as it collapses. As the cloud collapses, it loses gravitational potential energy, and as learned from our unfortunate accelerating whale, that energy has to go somewhere. When we force gas to occupy a smaller space, or fill a space with more gas molecules, we increase the pressure. We can do this on purpose to inflate a car or bike tyre, for example. When we do this, the molecules have less freedom to move, and they crash into each other and the walls of their container more – their thermal energy has increased. The gravitational potential energy has been converted into thermal energy, making the gas hotter and more energetic, moving about much more. The force, which we call pressure, provided by this increase in thermal energy acts in the opposite direction to the gravitational force, slowing down the contraction, and creating a tug-of-war between the two forces. When gravity is winning, the cloud contracts. When pressure is winning, the cloud expands, which can be catastrophic for our hopeful protostar. If you will allow me to return to the site where our unfortunate whale fell from the sky, we can observe the result of too much pressure. A beached whale is a sad affair but its disposal is serious business. Whales are rather large and their decomposition can cause a build-up of gas inside the carcass. As more and more gas accumulates, the pressure increases until the poor whale explodes, potentially causing 'a stinking mess. This blood and other stuff that blew out on the road is disgusting, and the smell is really awful', as one unfortunate whale-explosion witness reported to the BBC.[6]

The production of giants

The key to cooling down gas is collisions. Electrons around a nucleus occupy discrete energy levels and a colliding atom can trigger an electron to move up an energy level. Electrons are couch potatoes and much prefer to sit at the lowest energy level, so will descend back down at the first opportunity, emitting a photon of the equivalent energy drop as they go. In this way, the kinetic energy of a gas molecule is converted into radiative energy in the form of a photon that can more easily escape the cloud. In the primordial metal-free gas clouds, there wasn't much variety in the kinds of collisions that could occur. The hydrogen atoms could collide with each other and that was pretty much it. When two hydrogen atoms collided, they could bond to form molecular hydrogen. Molecular hydrogen is the state we observe most hydrogen on Earth and in the wider Universe. It can be created with the help of a 'catalyst' electron, which is captured by a hydrogen atom, producing a photon in the process. This hydrogen 'anion' can then bond with a regular hydrogen atom to make molecular hydrogen, releasing the catalyst electron, which can then go on to kindle more reactions.

The intermediate production of the photon plays a vital role, carrying away a packet of energy from the gas, and cooling it. This reaction happens readily in the hydrogen gas clouds, and the many photons produced result in a reduction in the thermal energy of the system, lowering its temperature and the internal pressure, and causing the cloud to collapse. In the primordial gas clouds, molecular hydrogen cooling resulted in the lowering of the temperature of the gas clouds to a few

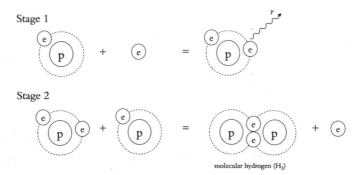

Figure 17 *Molecular hydrogen cooling. The formation of molecular hydrogen produces a photon (γ) in the process. This photon carries off energy, cooling the gas.*

hundreds of K. There are more efficient cooling avenues available to present-day gas clouds due to, for example, higher metal content, allowing modern gas clouds to reach much lower temperatures of only a few tens of K. The high temperature of the primordial gas clouds had significant consequences for the properties of the first stars. In modern-day star formation, the low temperatures of the gas clouds means that the Jeans mass is about 1 solar mass, so that most star formation results in stars about the size of our Sun. For the hot primordial gas clouds, however, the Jeans mass was up to hundreds of times larger, and gas clouds fragmented into much larger clumps, producing massive first stars. While similar size stars do exist around us today, they are rare. For the first stars, however, these immense masses were the norm. They lived on an entirely different scale.

Toxic stars

Most life 2.5 billion years ago, apart from the cyanobacteria, was anaerobic, so that it did not require oxygen to live or was even adversely affected by its presence. For anaerobic life the Great Oxygenation Event was in fact an extinction. The Great Oxygenation Event provided a double whammy. Not only did it kill everything, it also made life very difficult for anything that did survive. As the methane dropped from the atmosphere, so did the Earth's ability to trap sunlight. The Earth cooled and cooled, entering an ice age that would last almost 500 million years. It is even hypothesised that the Earth became a snowball,[7] covered by ice, or at the very least a 'slush ball', where the equatorial oceans were slush and the caps were covered in glaciers. Eventually the ice age broke, probably because underwater volcanic eruptions pumped ash and some greenhouse gases back into the atmosphere, increasing the retention of sunlight.[8] Cyanobacteria survived. Bacteria can be remarkably resilient. These days you can find them thriving in volcanic springs in Yellowstone National Park, under the ice of Antarctica and in the middle of the Atacama Desert. In comparison, British people can only leave their homes when temperatures are 5–25 °C, and the rain is no harder than a mizzle or drizzle. Due to the environmental pressure of the snowball Earth and the Great Oxygenation Event, only the most resilient life-forms survived – life-forms that would later become multicellular, and evolve in multiple directions, into creatures such as dinosaurs, hamsters and Mick Jagger.

Compared to cyanobacteria, the first stars were unlikely to survive their own environmental intervention.

The first stars had a surface temperature of about 100,000K (our Sun has a surface temperature of 5,800K) and this means that they produced a lot of energetic photons. The trouble is that most of the radiation was in the form of UV photons in what we call the Lyman-Werner band of energies: the exact energies that are able to excite the electrons a little too much in the molecular hydrogen, H_2. When the molecular hydrogen in the left-over gas surrounding the star absorbs these photons, it breaks up, destroying the molecule. Molecular hydrogen plays a vital role in cooling down the gas enough so that it can condense into a star. Without it, the cloud remains too hot as it tries to collapse, and expands instead, remaining a cloud of gas. In this way, then, the first stars were toxic to their own environment, destroying the material future stars forming from the primordial gas needed to cool down. This prevented further star formation in the surrounding regions. Molecular hydrogen was the key to the birth of the first stars and, for its efforts, it was destroyed.

They were stars with a mission to change the Universe for those that succeed them, to instigate a transition from simplicity to complexity. Before their intervention the Universe was dark: a hidden web of dark matter traced by invisible gas clouds. After it, the Universe was full of structures of different colours, shapes and sizes. There are red giants, blue giants, white dwarfs, brown dwarfs, spiral galaxies, black holes, humans and duck-billed platypuses. There are stars with little metal content (Population II) and stars of more metal content (Population I) ... but no obvious metal-free stars (Population III). The first stars don't seem to be around to enjoy the show they set the stage for. They were the architects and victims of their own mass extinction. Only when the first stars died and released their metals, enabling metal cooling, could the next generation be born.

For a long time we thought this meant that the first stars were lonely stars: once one had formed, the star-forming region was quickly destroyed so that no siblings could form. It's an enigmatic idea – a dark, empty Universe with isolated stars blossoming into life. The idea of lonely stars is losing credibility, however.

Stellar siblings
In present-day star formation we can watch stars form in nurseries. Big clouds of gas sub-fragment into stellar-sized

collapsing clumps, producing families of stars in close proximity: a stellar nursery. This is because the gas is cool, the Jeans mass is low, so that even less massive clumps within the cloud can become gravitationally bound and collapse. Due to the unique metal-free composition of the very early Universe, it was only the large gas clouds that were able to collapse at all. The clouds were too hot due to a lack of available metals to cool them down enough, so larger masses were required to fight back against the radiation pressure. A few years ago we thought star formation was singular for Population III stars. They formed alone, lived alone and died alone. Except that when we got ever more advanced computers, we could apply the same level of detailed chemical reactions and gravitational interactions not just for the clump at the centre, but for the whole gas cloud.[9]

First, our imaginary symmetrical spherical gas cloud must get a reality check. When these dark matter halos form, and the gas halos within them, they interact gravitationally with any surrounding halos. This imparts a torque: a rotational velocity of the cloud that might be tiny at first, but as the cloud contracts this spin can increase quickly, just as pirouetting figure skaters spin faster as they bring in their arms. This increased rotational velocity forces the gas cloud into a more flattened disk shape that spirals in to the centre where the first protostar forms. This is called an accretion disk. Over time, gas moves across this accretion disk, accreting on to the main protostar and making it larger and larger. At first the star is small, about 1 per cent of the mass of the Sun, but with time it accretes so much hydrogen that it swells to about 100 times the radius of the Sun. The rate of

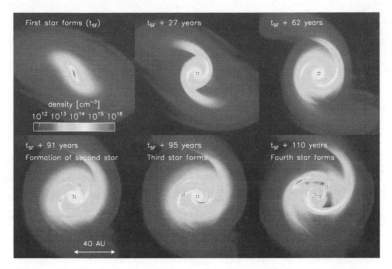

Figure 18 The formation of Population III stars. In these six simulation snapshots we see a disk of gas fragmenting into four protostars, in a short timescale of just over 100 years. 1 AU is the distance between the Earth and Sun. ©Science Clark et al. 2011.[9]

accretion on to the star is so large that the flow compresses the volume down instead of growing the physical size. This contraction continues until the star has grown to be 100 solar masses or more, yet with a radius only a few times as large as the Sun.[10] This was the bit we've known for a while, but it is only recently that we've been able to zoom out and follow what is happening to the rest of the accretion disk. In Figure 18 we can follow the formation of not one but four protostars over a very human timescale of only 110 years. What we are seeing is the fragmentation of the accretion disk. We think this happens due to a variety of factors, for example turbulence in the disk resulting in local patches of lower Jeans mass. The speed at which a star accretes mass is proportional to its mass, and the first central protostars are undoubtedly massive. The mass is falling on to the disk with such a speed that it cannot travel inwards fast enough. The disk is overwhelmed with mass, compressing into a higher density and decreasing the Jeans mass, and local areas of the disk can collapse shortly after the main protostar forms. There have been multiple simulations now, all showing the same thing: that the first stars are anything but lonely.[11,12,13] Some predict one to a few siblings, others upwards of a hundred. The masses of these stars vary over a wide range from just under a solar mass to 1,000 solar masses. The 'main' first stars in these simulations were massive, and they shut off star formation quickly – but not quickly enough to prevent a family forming around them first.

The first stars may not have formed alone, but they may not have been 'happy families'. The smaller protostars don't necessarily stay within the accretion disk. Some will stay as separate sibling stars. Some will merge with the protostar, adding all its gravitationally bound hydrogen to the fuel source of the first star. Others, perhaps as many as half, will get ejected,[14] their orbital velocities too large to remain bound in the halo. These lower mass, ejected siblings could provide us with a way to look back into this time by acting as living fossils, safe from the obliteration that eventually comes when living next to a gigantic star.

The Cosmological Oxygenation Event

Once formed, these first stars lived brief lives, by the very nature of their large masses. As they tried to combat the huge gravitational pressure, they burned through their fuel extremely quickly, producing heavy metals along the way. The hydrogen began fusing in the star's core, and over its lifetime it burned more and more hydrogen, in shells of ever-greater radius. Below the hydrogen burning shell, the next level of fusion could proceed: helium burning. Once there is helium, more reactions can start, producing carbon and beryllium, neon and oxygen. This proceeds along the periodic table, with heavier and heavier metals created in onion-like shells. The result of all this fusion is to produce the energy and pressure required to hold up the star against the relentless crushing force of gravity. But as the star runs out of hydrogen and progresses to burn heavier elements, it becomes harder and harder to produce the resources needed to push back. The heavier elements release less net energy per unit mass of fuel, so to maintain the same energy holding up the star, the fuel is used up more quickly. For a star 20 times the mass of our Sun, while hydrogen burning in the core would last 10 million years, helium burning would last 1 million years.[15] For carbon burning it would then burn for 300 years, and oxygen would burn for 200 days. Silicon burning would last only two days, and once we reach iron, the fusion engine judders to a stop. The energies that enable iron to be fused also enable photons to be so energetic that they are able to destroy iron and other heavy nuclei. The heavy elements break up into the constituent protons and neutrons, and the pressure of the star collapses. Once the star begins to fuse iron, the star can no longer produce the same amount of energy, and the relentless push of gravity finally overcomes it. The star's structure is ripped apart and flung, metals and all, into the surrounding environment: a Cosmological Oxygenation Event if you will.

Within their formation halo the stars expel their metals, muddying the pristine primordial gas in a way that cannot be undone. The pristine gas is remarkably prone to pollution. As the first generation of stars dies and pollutes its surroundings, the metal fraction of the gas increases from zero to 1/1000th that of the Sun today.[16] This is enough to mark the transition between Population III to Population II star formation. The first stars were made from a very basic set of ingredients: no vanilla extract or cocoa powder, no icing on the cake or cherry on the top. It is the composition of the stars that marks the difference between a first star and a present-day star. As soon as they die, the gas is enriched enough so that no more Population III stars can form within that halo. Now this hot, enriched gas must cool, and settle again into lucky and unlucky clouds of gas until they form a new generation, Population II. This happens quickly, cosmologically, only 10–100 million years after the Population III supernova. The seeding of the Universe with metals happened in the blink of an eye cosmologically, as did the Great Oxygenation Event geologically.

Islands in the sky
Some supernovae explosions are powerful enough to expel metals out of the formation halo into the wider intergalactic medium (IGM), but they soon fall back in again because of the gravitational pull of their home town, like millennials to their parents' spare rooms. The explosions of the first stars are powerful, but they are unsuccessful in changing anything but their local environment. Due to this localisation of metal ejection,

that metal enrichment of the intergalactic medium is patchy:[17] it follows the distribution of halos, which is not smooth across the sky. Halos form at the junctions between dark-matter web filaments, and in simulations you can see how the metal enrichment follows the web.* If the supernovae were more energetic, then we would expect to see a homogeneous spread of metals very quickly, not a sprawling of metals tracing the web as we do. The sprawling increases with time, in a fashion reminiscent of bacteria in a petri dish. This patchy pollution opens up a possibility. If the metal enrichment of the IGM is patchy, then the lowest density voids in the cosmic web would be the islands in the sky, safe from the growth of metal-contaminated gas around them. In these havens, pristine gas could be preserved, creating Population III stars up to a much later date because of the slow coalescing of low-density gas. We are talking about such a long time ago, dare I repeat it again, over 13 billion years, that it is a very long way to look with a telescope to see light that old. Any process that brings these stars closer in time, and therefore distance, will result in more chance of seeing these first stars. If we can focus our telescopes on areas of low density, voids betrayed by their lack of large structures, we might find evidence of metal-free stars much closer in distance than we would in higher density patches.

★ ★ ★

---

*You can watch all these inter-playing quantities, dark matter, gas, temperature, metal enrichment fraction play on the Illustris Project website.

The first stars were different from the stars of today both in nature and in the way they formed. The lack of cooling mechanisms beyond molecular hydrogen cooling resulted in the hot primordial gas clouds having a large Jeans mass so that only large clumps of gas could collapse. We used to think that this meant that the first stars formed in units of one, but due to improved simulation methods we can now follow what happens in the surrounding accretion disk. Now we believe that the first clouds collapse into a massive central star, with smaller protostars forming around it. Some are ejected, some are absorbed into the central mass and some survive, producing first star binaries. This first burst of star formation quickly prevents a second generation of Population III stars, as the molecular hydrogen necessary for cooling is destroyed by the photons from the first stars themselves. They live their short lives, fusing metals and eventually expelling them into their immediate surroundings with such aplomb that any stars forming from that gas already belong to Population II. In the blink of a cosmological eye, the massive first stars are gone – but their legacy remains, the seeding of the Universe with metals.

In Shark Bay,[*] in Western Australia, within an area called Hamelin Pool,[†] cyanobacteria have formed

---

[*] In case you are wondering, Shark Bay is not known for its sharks. The tourist website has a typically Australian non-warning: 'Sharks are no more a threat in Shark Bay than anywhere else around Australia. Of the 28 shark species recorded in Shark Bay only one or two may arguably be considered dangerous.' Oh, that's all right then. Just one or two dangerous sharks.
[†] Eighty-five Windermeres.

bacterial mats: sheets of bacteria matted together. A long time ago, before the Great Oxygenation Event and for some time after, these mats were thought to be ubiquitous. Once grazing animals evolved, the mats became lunch and today only a few locations host them – locations which in one way or another are protected from grazing animals. As the water washes over the mats, a fine layer of sediment floats down, glued by the sticky bacteria on to the thousands of layers that have come before. Over hundreds of years these living rocks get bigger and bigger, growing at about 0.3mm a year, to form living fossils such as those that protrude in Shark Bay. Older examples of these stromatolites are some of the oldest fossils on Earth. In Glacier National Park in Montana, US, you can view stromatolites from more than a billion years ago, 'sliced cabbage' rocks that give us a record of some of the earliest and most resilient life-forms. These fossils are not always easy to find, and have been exposed by human intervention or geological events. They lay there camouflaged by other sediments. If ejected or late-forming Population III stars are to be found wandering around or adopted by other stellar systems, then they are most probably camouflaged by the metals they encountered on their travels. Uncovering them is no easy task and requires some digging.

CHAPTER SEVEN

Stellar Archaeology

I never wanted to be an astrophysicist. While a lot of my colleagues were looking through amateur telescopes, I was dreaming of decoding hieroglyphics and brushing off hidden artefacts in newly discovered Ancient Egyptian tombs. I tried (and failed) to learn the language and spent my work experience packaging 5,000-year-old Egyptian shoes at a local museum on Tuesday afternoons. At 16, a curator had warned me that the chances of getting a job in Egyptology were vanishingly small, and as I looked around the library I turned away from ancient history to a book on time travel on a nearby shelf. I remember thinking that the book had been misplaced in the non-fiction section, but I took it home all the same. I read all about Einstein's special relativity: how time slows down the faster you move and how a ruler appears shorter if it moves at speeds approaching the speed of light. I was astonished that I hadn't heard of this before – and didn't understand a word of it. In only one sitting, it was enough to pull me away from my pursuit of the past into something that sounded like science fiction. I went to university to study physics just so I could understand that book. It's quite fitting for me that even within physics I then made it about the past. I could have made unhackable quantum computers or levitated trains to make them go faster.

I could have researched teleportation, invisibility cloaks or artificial intelligence, but in the end my passion remained in the opening of an undiscovered tomb, this time on a cosmic scale.

As is the case with most young Egyptophiles, there is one story that captured the excitement of Ancient Egyptian discoveries more than any other, and that is the discovery of the tomb of Tutankhamun. In November 1922, British archaeologist Howard Carter held up a candle and peered through a small drill hole in the tomb door. His patron, Lord Carnarvon, asked him if he could see anything, to which Carter, struck dumb with amazement, could only reply, 'Yes, wonderful things, wonderful things!'[1] He recounted later that, in the dim candlelight, 'details of the room emerged slowly from the mist ... strange animals, statues and gold – everywhere, the glint of gold'. Everywhere, the glint of gold. My heart still races when I read that story. Imagine that moment – drilling through a door that had not been opened in more than 3,000 years and wondering what you would find. I feel the same way now, looking for the first stars, searching for anything they might have left behind – even some discarded shoes would do.*

The unveiling of Tutankhamun's final resting place was exceptional because of how rare an occurrence it was. Most times a tomb was 'discovered', grave robbers had discovered it first, sometimes thousands of years

---

* Interestingly, there is an astronomical connection in the story too. An iron blade was found resting on Tutankhamun's right thigh. Recent analysis has shown that the iron originated from a meteorite, similar to a few of the rare Ancient Egyptian iron items found.[2]

Left: The Sun's corona as it appeared during the 2017 total solar eclipse.

Left: The corona as viewed during the 1925 eclipse, by artist Howard Butler.

Below: The Centaurus A galaxy as observed in only in the optical wavelength and with the vast radio lobes superposed.

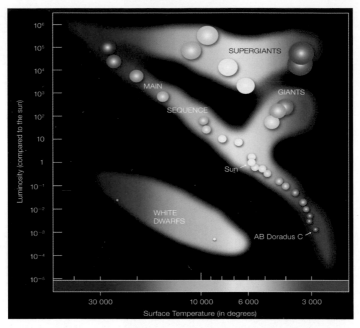

Above: The Hertzsprung-Russell diagram.

Below: The Andromeda galaxy, one of our closest neighbours.

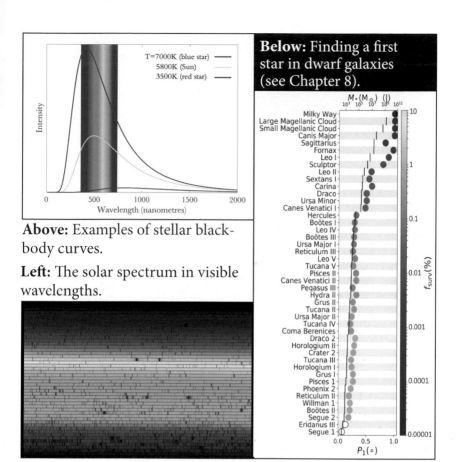

T=7000K (blue star)
5800K (Sun)
3500K (red star)

Intensity

0   500   1000   1500   2000
Wavelength (nanometres)

Above: Examples of stellar black-body curves.

Left: The solar spectrum in visible wavelengths.

Below: Finding a first star in dwarf galaxies (see Chapter 8).

$M_*(M_\odot)$ (l)
10^3 10^5 10^7 10^9 10^{11}

Milky Way
Large Magellanic Cloud
Small Magellanic Cloud
Canis Major
Sagittarius
Fornax
Leo I
Sculptor
Leo II
Sextans I
Carina
Draco
Ursa Minor
Canes Venatici I
Hercules
Boötes I
Leo IV
Boötes III
Ursa Major I
Reticulum III
Leo V
Tucana V
Pisces II
Canes Venatici II
Pegasus III
Hydra II
Grus II
Tucana II
Ursa Major II
Tucana IV
Coma Berenices
Draco 2
Horologium II
Crater 2
Tucana III
Horologium I
Grus I
Pisces 1
Phoenix 2
Reticulum II
Willman 1
Boötes II
Segue 2
Eridanus III
Segue 1

f_{surv}(%)

10
1
0.1
0.01
0.001
0.0001
0.00001

0.0   0.5   1.0
$P_1(\circ)$

Below: In the Harvard stellar classification system, an O star has an average surface temperature of 30,000K or above, while an M star will be closer to 3,000K.

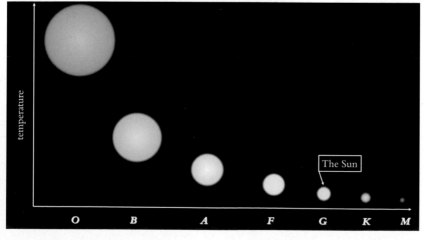

temperature

The Sun

O   B   A   F   G   K   M

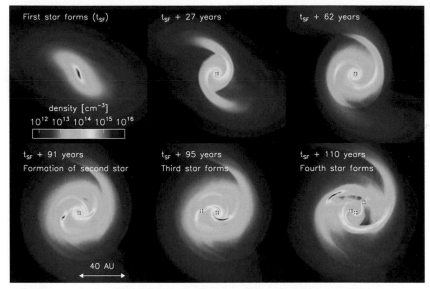

Above: A disk of gas fragments into four Population III protostars in a short timescale of just over 100 years.

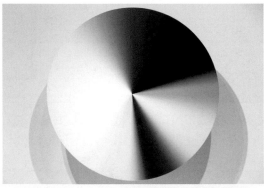

Left: 'The Cosmic Spectrum' by Katie Paterson. Photo by Manu Palomeque.

Left: Cecilia Payne-Gaposchkin.

Above: Penzias and Wilson stare at the Holmdel Horn Antenna at Bell Telephone Laboratories.

Right: The Planck cosmic microwave background map.

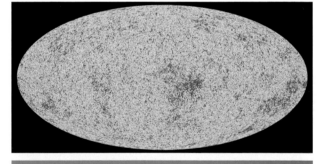

Right: Stromatolites in Shark Bay, Western Australia.

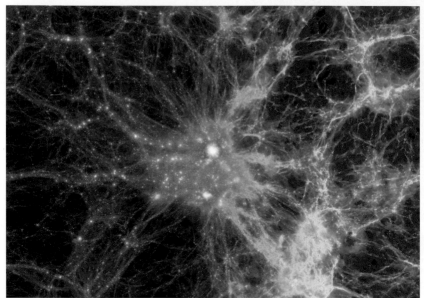

Above: The Cosmic Web. The simulated dark matter distribution (left) and gas distribution (right) from the Illustris cosmological galaxy formation simulation.

Below: The Hubble Ultra Deep Field.

Left: The event horizon of M87's supermassive black hole, as imaged by the Event Horizon Telescope.

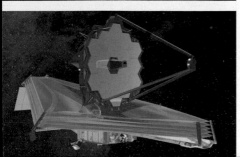

Left: The Mice galaxies are in the lengthy process of colliding and merging.

Left: In the Antennae galaxies, a violent merger has ignited fresh rounds of star formation.

Left: An artist's impression of the James Webb Space Telescope.

Above: The Aperture Array Verification System, the testbed for potential SKA technology, with the Milky Way overhead.

Left: Here, I am very proud of a LOFAR station in the 'superterp' or central core of the radio telescope close to Exloo, Netherlands.

Below: The Green Bank Telescope on 15 November 1988 (left) and 16 November 1988 (right).

before. Ancient Egyptians realised pretty quickly that if you mark the spot where you buried the treasure with a gigantic pyramid, the treasure would not remain buried for long. They began to secrete their kings, queens and the accompanying treasure underground, but even that wasn't enough to keep their treasures safe, making Tutankhamun a rare exception. Much of Egyptology is actually based around piecing together fragments of knowledge: half a sarcophagus here, a mummy there, a canopic jar found in an attic. It's amazing how the detritus of daily life can be found in strange places, gathering dust for thousands of years before we see it for what it is. Here we can draw a parallel with our search for the first stars, because the Universe has changed so much since those first stars were laid to rest. The pristine environments in which they formed were raided, polluted by interloping young stars and their messy supernovae, leaving little scope for undiscovered stellar tombs. Somewhere, however, there might just be some artefacts from 13 billion years ago hanging around our local neighbourhood, and to uncover them we need to do some stellar archaeology.

The initial mass function
It shouldn't be so hard. After all, the glint of starlight is hard to miss and we certainly aren't short of places to search in the Milky Way. The prevalence of stars presents a problem, though, and searching for a metal-free star among the billions of Milky Way stars is like searching for genuine Ancient Egyptian artefacts at a jumble sale in Texas. You might get lucky – but you'll probably end up walking away dejected, wearing a battered

second-hand cowboy hat. One of the most useful
relations in stellar studies is the mass-lifetime relation.*
Stars have different lifetimes depending on their mass,
with smaller mass stars having less fuel but using it up
slowly, rationing it out for billions of years. More
massive stars have more hydrogen to use as a fuel for
fusion, but they use it at such a high rate that the fuel
will run out within a few million years. We have seen
from simulations that the first stars could have been a
hundred or perhaps even a thousand times the mass of
our Sun. A quick consideration of the mass-lifetime
relation tells us that we therefore expect the first massive
stars to live short lifetimes of only a few million years, a
hopeless situation for those of us hoping to observe one
13 billion years later.

But don't worry, all is not lost. When we have been
talking about our expectations of Population III stars, we
have been speaking implicitly about our expectation of
an 'average' Population III star. We think now that the
first stars formed in nurseries, potentially with lower
mass companions. Overall, yes, we expect the first stars
to have been more massive than your average star today,
but as with every measurement we also expect there to
be a distribution around that average. If you have ever
bought clothes for a child, you have probably noticed
that they are sold in sizes such as 'Age 9, 132cm'. This is
not as a result of every single nine-year-old in the world

---

* In science a relation is not an annoying person you have to make
polite conversation with every Christmas, but a connection
between two properties. For example, the relation between time
spent with relatives and stress felt would be linear ... or perhaps
exponential..

being precisely 132cm for 365 days, but because the average nine-year-old will measure that height. If we consider the actual heights of a population of 1,000 nine-year-olds, there is a spread of values around that average: a small number are much shorter than average and a small number are much taller than average[3].

These height distributions differ across different countries, different ages and different decades. For example, in North America, the average height of a human male increased from 1.68m (5ft 7in) in 1896 to 1.76m (5ft 10in) in 1996,[4] which is why when visiting historic houses you are likely to need to mind your head.[*] We can easily form these height distributions, or height functions, back in time as long as we have data, but at some point humans were just too busy finding food and fighting the plague to care about what height they were, let alone record it. We can still reconstruct height distributions from earlier times, however, by measuring a large enough sample of human remains. By measuring the heights of mummies found in Ancient Egypt, we can even come up with what we might call an Initial Height Function, a height function of one of the earliest societies. For example, recent research has

---

[*] You might be wondering if this relationship between time and average height is a linear one, i.e. will humans be even taller in another 100 years? The answer is probably no. The average height in the US today is much the same as it was in 1996 and we've begun to see a plateau in the height increase in recent decades, indicating that we have reached our genetic limits in the environmental conditions we live in now. Even the Dutch have slowed down, after rapidly gaining a whopping 13cm (9in) in 100 years to become the tallest men on Earth at an average 183cm (6ft!).[5]

suggested that royal Ancient Egyptian males were on average taller than the general population, and that royal women were on average shorter, suggesting a shared genetic characteristic and adding weight to theories of inbreeding within the royal family.[6] Tutankhamun died aged 19 at an approximate height of 1.8m (5ft 11in), which is on the tall side for his age, even by modern standards, leading the same research to suggest that he may have been the product of a sibling union. Like historians, stellar archaeologists must look at the indirect evidence around them to constrain the initial mass function (IMF) of Population III stars. If we can find out how many stars of each mass were created, then we can determine if there is any hope of finding a low mass, ancient Population III star still around today. Looking at the mass–lifetime stellar relation, we can estimate that only stars 80 per cent the mass of our Sun or less can have survived to the present day, so the golden question for stellar archaeologists is: did the Population III IMF go that low?

When it comes to nearby groups of stars, we can construct stellar mass functions with relative ease, given the right technology: we simply count the number of stars we observe at each mass. With the first stars, the job is harder because there is no sample to draw from. We've never seen a Population III star, or even a remnant of one, so we do not even have the bones of an ancient civilisation from which to draw our conclusions. Instead, we use simulations with as much physics and early Universe knowledge as possible to consider all the potential scenarios. What we find is that the Population III IMF is top heavy:[7] most of the mass in a stellar sample is found in high-mass stars. The usual mass functions of the stars we observe around us are bottom heavy, as we observe many more low-mass stars than high-mass ones. This is partly because a lot of the older high-mass stars are dead, and partly because the conditions of the Universe today favour the formation of lower mass stars in the first place. In contrast, the early Universe favoured star formation on the higher end mass scale, due to the lack of cooling metals. Population III stars were massive compared to what we see around us: they were the male royals (or perhaps the Dutch) of the sky. Critically, however, when we simulate the possible Population III IMFs, while there are many stars on the higher mass scale, there is a tail in the distribution where a few stars are created at low enough masses to still be observed today.[8] There is a lot of uncertainty around Population III IMFs, but the possibility of living fossils in our neighbourhood is there, and that's enough for us to search.

Even searching and failing is a result in science. Imagine a closed box, filled with 100 balls coloured red, yellow and blue and, hey, why don't we make this interesting and make every red ball grabbed from the box worth £1,000. Perhaps red balls make up 80 per cent of the ball population, perhaps 25 per cent, or perhaps there are none at all. If we close our eyes and grab a ball and it isn't red, that doesn't tell us much as there are plenty more places where the red balls could be hiding and most of our models (apart from the 100 per cent red one) are still in play. If, however, we pick out ball after ball and none are red, we become increasingly dejected as more of the viable models with red balls are ruled out – we are constraining our model space. This thought experiment is not a perfect analogy, as grabbing a ball from a box is a simple manoeuvre, whereas searching for a metal-free star is not. For a start, the box (Universe) is much, much bigger, so that our telescopic arm cannot even reach a lot of the balls and, as we will find out, those it can reach are less likely to be red than those further away. Every grab you make also costs time and money – both quantities that are prized and protected in academia and certainly don't allow for infinite attempts. In addition, determining whether a Population III star is metal-free and not just metal deficient is complex in itself – the balls have been in there a long time, all mixed up, faded from use, their colour hidden by dirt. Is it a red ball we hold in our hand, or a blue one? It's hard to tell, and we need a way to be sure.

An abundance of metals

To the human eye, an 80 per cent solar mass star born 13 billion years ago looks much the same as an 80 per cent mass star born 4 billion years ago. Their stellar spectra are very different, though. The multiple iron absorption lines, caused by the many atomic energy levels in an iron atom, are much weaker in the more metal-poor star, because there is less iron. We assess the metal content by measuring what fraction of the star comprises a particular metal based on the strength of the metal absorption lines in the spectrum and comparing that fraction to the same quantity in the Sun. An 'iron-poor' star is composed of a lower fraction of iron than the Sun, even if the star is very large and has a larger amount of iron in terms of mass or atoms. There are lots of metals to choose from, but iron produces strong absorption lines, which are most easily measured by optical telescopes, and it turns out that iron is a good indicator of overall metallicity too. In most stars, and we'll talk about the exceptions later, if we measure a low iron fraction we can use 'iron-poor' and 'metal-poor' interchangeably. And so, generally, if a star comprises a smaller fraction, or *abundance*, of iron than the Sun, then we call it *metal-poor*, and anything with a larger fraction of iron is called *metal-rich*. The stars we see around us are metal-rich because they formed from gas already seeded with metals by previous supernovae. The actual fraction of iron in the Sun is about 0.14 per cent by mass[9]. Stars forming today typically have an overall metal content of about 2 per cent by mass;[10] only 2 per cent, yet we call them metal-rich! It's worth remembering that hydrogen will always be the main dish in a star by far. The difference between

NAME	DEFINITION
The Sun	solar abundance
Metal-poor	1/10th solar iron abundance
Very metal-poor	1/100th solar iron abundance
Extremely metal-poor	1/1,000th solar iron abundance
Ultra metal-poor	1/10,000th solar iron abundance
Hyper metal-poor	1/100,000th solar iron abundance
Mega metal-poor	1/1,000,000th solar iron abundance
Septa metal-poor	1/10,000,000th solar iron abundance
Octa metal-poor	1/100,000,000th solar iron abundance
Giga metal-poor	1/1,000,000,000th solar iron abundance
Ridiculously metal-poor	1/10,000,000,000th solar iron abundance
Supercalifragilisticexpialidociously metal-poor	Less than 1/10,000,000,000th solar iron abundance

Figure 19 *The current nomenclature of stellar metallicity,*[*] *adapted from Frebel 2018.*[11]

---

[*] *Okay, I made the last line up, but I think it is a strong contender.*

a Population III star and a star today is merely garnish, yet it makes all the difference (as does the parsley on your potato, and the basil on your pizza). As we search for older and older generations of stars, the already low metal content decreases until we are dealing with extremely low metal fractions that require accompanying superlative labels[11] (Figure 19). In fact, stellar archaeologists rarely get out of bed for any star unless it has less than 1/10,000th the iron fraction of the Sun: *ultra metal-poor* stars.

The discovery of metal-poor stars

The first metal-poor stars were discovered in 1951, by American astronomer Joseph Chamberlain and his PhD supervisor Lawrence Aller,* who measured the iron abundance in two stars and found they were about 1/10th the solar value.[13] This may seem a small difference from our Sun, but at the time it was revolutionary, and Chamberlain struck gold in the academic sense, too, as this was his first ever paper and it turned out to be one the most important astronomical papers of the twentieth century. Until that point it had been assumed that all stars had roughly the same elemental make-up as the Sun. Stars with different strengths of metal lines had been observed, but this was explained away by the stars being of different spectral type only, and not indicative of a different generation entirely. Spectra were, after all,

---

* Aller is interesting in that he became a professor of astrophysics despite never having completed high school. He started high school but was dragged away by his father to a gold-mining camp.[12] He managed to get away and went mining for metals in stars instead.

the tool that was used to separate out the stellar zoo into the OBAFGKM classification system in the first place, with the strength of the hydrogen lines becoming a proxy for temperature. In Chamberlain and Aller's paper you can feel their disbelief as they reveal their results, book-ended with phrases such as 'the observed [*iron*] abundance appears to be smaller than in the sun, although this conclusion must be taken with caution'. They refer to the measured iron abundances as 'the one undesirable factor in our interpretation'. Another case, then, of the undesirable noise becoming the paradigm-shattering signal, though with less pigeon poo this time.

It's easy to look back in hindsight and cherry-pick these kinds of statements from the unassuming papers that changed our world view, but what we are looking at are good scientists. They had a sample size of two metal-poor stars against the accepted knowledge that stars were all the same, all like our Sun, and while a scientist should be confident in their analysis to publish, rarely does a eureka moment come so clearly. More often, instead of 'Eureka! I've got it!', it is 'That looks weird! Hang on, what is that? I think that might be … huh. Hey, can you all check this to make sure I haven't done something stupid?' It often requires a few eureka moments for people to be comfortable leaving their comfy, but wrong, theories. When you read backwards from Chamberlain and Aller's work you see hints that the ground was already shifting under the homogenous star theory, for example in their reference to American astronomer Nancy Roman's work the year before. Nancy Roman had a major role in planning the Hubble Space Telescope and was the first woman in an executive position at the

National Aeronautics and Space Administration (NASA), but before all that her work provided a hint that stars were not all the same. What she did was take a sample of stars and categorise them firstly by how weak their metal lines were, and secondly by their velocities.[14] As far back as 1926, Dutch astronomer Jan Oort had noticed that the Hertzsprung-Russell (H-R) diagrams of high-velocity stars were markedly different from those of lower velocity stars. This difference was confirmed by Baade, who would observe high- and low-velocity stars in Andromeda from Mt Wilson Observatory in 1944, when the wartime blackouts allowed.[15] Exploring the different velocity populations further, what Roman noticed was that when she divided the sample into weak and normal metal lines, it was only in the weak group that she saw the high velocities. Roman had discovered that stars with more metals tended to be found dawdling about in the disk of the Milky Way. In contrast, more metal-poor stars tended to be found in high-velocity elliptical paths that extrapolated to them residing often all the way out in the halo. It would take Chamberlain and Aller to pin down that weak lines did indeed mean metal-poor, and not just different, spectral type.

Searching the Galaxy

There are plenty of stars similar to the original Chamberlain and Aller stars, 1/10th the solar iron fraction, but the push down to the iron levels that are of interest for stellar archaeology has been far more challenging than hoped, because the search area is just too big. But we shouldn't lose hope – nothing worth having comes easy, as they say, and it is never mentioned that Howard

Carter didn't just stumble upon Tutankhamun's tomb. In 1914 Carter and Carnarvon bought the site permit to dig from an archaeologist who was convinced that he had found everything there was to find there,[*] and due to the outbreak of the First World War, Carter didn't start work until three years later. Even then he didn't just go at it with a trowel, but instead divided the archaeological site into a grid and spent five years systematically and meticulously exploring each square before he finally got lucky. Imagine that diary: day 1: nothing, day 2: nothing, day 1,824: nothing, day 1,825: thousands of artefacts, untouched mummy of the lost boy king. Beans on toast for tea.[†] If you ask the Internet how many stars there are in the Milky Way it says there are 250 billion – plus or minus 150 billion. That is a big error bar by all accounts. Counting the number of stars in the Milky Way is hard because of the large volume of faint stars that we cannot detect. Because even the brightest stars across the Galaxy appear very faint to us, it's something like trying to predict the population on Earth when you only have the census figures for Milton Keynes. Even if we go to the lower end of that estimate, 100 billion stars, that is still an awful lot of stars to sift through and determine metallicities for. If we had a dedicated telescope (we don't), and could

---

[*] That's right up there with the man who turned down The Beatles and the editor who rejected J. K. Rowling.

[†] Carnarvon had about had it with the lack of return on his investment in Carter and pulled his funding shortly before the discovery. Carter successfully begged one more season and four days into the dig found the biggest archaeological discovery in history. Which just goes to show that even lost mummies are found in the last place you look.

analyse one star per second (we can't), it would still take more than 3,000 years to check them all, so we need to narrow the search area.

There are three luminous components of the Milky Way: the disk, the bulge and the halo. The disk is a large, pancake-like structure containing the spiral arm structure. It is the site of active star formation and thus the home of mostly young Population I stars, so that's a no-go: you might find the odd artefact in a tourist shop ... but it's not a good bet. We think that around one in 200,000 stars are metal-poor in the solar neighbourhood of the disk.[16] The bulge is the dense, spheroidal collection of stars towards the centre of our Galaxy. Galaxy-formation simulations indicate that the bulge is the oldest component of our Universe, having formed either during or shortly after the initial collapse, so on paper it sounds like a brilliant place to dig. In reality, however, the bulge is a promising dig site but with extremely challenging conditions: think Indiana Jones jumping over pits of snakes difficult. Because it is the oldest component, it has been well seeded with metals by supernovae, and thus contains many high-metallicity stars to distract us from the few low-mass Population III stars that might remain. Worse, all these supernovae produce a prodigious amount of dust, which confounds our telescope trying to peer through the murk. The bulge may thus be a sensible place to look at theoretically, but technologically it is tricky, which of course isn't to say scientists aren't trying.[17] And so our hopes land with the halo. This is the more nebulous luminous structure surrounding the disk and bulge, and contains groups of stars known as globular clusters, and field stars that are not associated with any individual cluster. You might make the connection from Roman's work that lower metallicity stars move faster, which would be a rather handy identifying attribute. You could simply measure the speed of the stars and target the fastest ones. Sadly, there is no such causation alongside the correlation: low-metallicity stars are not by nature faster. Why are higher velocity stars more likely to have weak metal lines in their spectra? Star formation occurs mostly in the spiral, but given time young stars can move out of the disk plane, into the halo. This has very little rotation compared to the disk, so it is in fact us within our place in the disk who are moving at particularly high velocity. From our point of view, therefore, the stars in the halo are moving in the other direction at high speed, though they may in addition have a component of velocity perpendicular to the Galactic plane. In reality, these so-called 'high-velocity' stars are in fact just stars that orbit around the Galactic centre in a different plane. Really old metal-poor stars are more likely to have had the time to drift out of the stellar disk and into the sedate confines of the halo where, due to our own high rotation speed, they appear to be moving at high velocities, as observed by Nancy Roman. The halo thus seems an excellent place to search for Population III stars, and indeed that is where most of the effort in stellar archaeology so far has been concentrated, with outstanding success.

Stellar archaeology

As of mid-2019, only a handful of stars have been observed with an iron abundance below 1/10,000th.

Technical limitations with a telescope or interference from other stars introduce convincing forgeries into our spectral lines, and the analysis carried out by stellar archaeologists to understand and remove this contamination is vital work. The current record holder for the most iron-poor star where an iron measurement has been made is SMSS J1605-1443.[18] Discovered in 2018, this mega metal-poor halo star has less than 1/1,000,000th the solar iron fraction. This is exceedingly low, so is this our Population III star? Unfortunately, no, the levels of other heavy metals present in the spectrum are too high for this star to have created them all by itself – it has to have had help from a previous supernova seeding the cloud it is made from with a starter kit of those metals. It's the equivalent of faking an Egyptian mummy almost perfectly, but leaving the body with a smartwatch on. You probably haven't heard of the Persian princess, a mummy found for sale on the black market at the turn of the new millenium. For six months this mummy caused an international furore as Pakistan, Iran and apparently even the Afghan Taliban all staked a claim on what appeared to be a 2,600-year-old artefact worth millions. She was placed on display in the National Museum of Pakistan, but her vacation in fame was short. Oddities about how she was mummified, for example, the continuing presence of her heart, and the incorrect grammar on the sarcophagus, alerted an American archaeologist to dig a little deeper. What he found was not the 2,600-year-old mummy of a princess but the brutally murdered body of a 26-year-old woman who died around 1996. I tried to find out who that poor woman was, but as far I can tell she was never claimed and no one was ever charged with her murder, which is why I spend most of my time looking at the stars. Things down here are all wrong. Anyway, whether we are talking about stars or artefacts, there are plenty of forgeries about.

The levels of iron detected in SMSS J1605-1443 were low, but they were definitely there. There is one star where no iron has been detected at all and only an upper limit could be placed, SM0313-6708.[19] Observed in 2012, high-resolution spectroscopy of SM0313-6708 in 2013 revealed … nothing much at all. Instead of a forest of absorption lines, this star revealed almost no activity at all and the presence of only four metals (lithium, carbon, magnesium and calcium). Everywhere we should have seen a dip relating to the presence of iron, there was just a flat line, indicating an absence of iron. With our technology we cannot be absolutely certain that there isn't a tiny dip hiding among the general noise of the signal. But we are sure that we are looking at a star at least 1/10,000,000th more iron-poor than the Sun, and it could be far lower. This sounds pretty promising. We found a star that appears to have a total absence of iron, so surely this is our Population III star? Well, unfortunately, it's another no. Only four metal elements may have been detected, but still the abundances found were large enough that they couldn't have been produced purely by nuclear fusion within a first star. But it's agonisingly close. The metal levels detected in SM0313-6708 are so low that models show that it could have formed from a cloud enriched by just a single supernova. So what we are looking at here is not a Population III star, but could well be a first descendent. We can learn a lot about the first stars from a first descendent. From the metal levels

measured in this star, the authors could determine that
the likely progenitor was a 60 solar-mass star, which
exploded in a low-energy supernova before settling
down into a black hole. This model fits so well because
of the presence of metals like carbon and yet an absence
of iron. When a star ends its life in a low-energy
supernova, what we call a core-collapse supernova, then
only the lighter elements such as carbon and magnesium
are expelled in the outer layers of the explosion. This
leaves the heavier metals, such as iron, much closer to the
central black hole – and being close to a black hole is
never a good thing. We think, then, that while these first
stars produced iron, all that hard work was lost down a
black hole, with the second generation forming from
the lighter metals that escaped the black hole, but
not the heavier ones such as iron that we had been using
to track metallicity overall. The most recent metal-poor
star, SMSS J1605-1443, you remember, the lowest
metallicity star with an iron measurement, also fits this
core-collapse model, and this pattern of low iron
abundance and high carbon abundance is repeated in
all but one of the known ultra metal-poor stars. These
low-energy supernovae could be more common than
previously thought, when it seemed that all the first stars
ended their lives as much more explosive supernovae. If
true, this has significant consequences for how long that
earliest generation was able to exist as a whole. Low-
energy supernovae pollute the Universe inefficiently,
preserving pockets of metal-free environment, so
Population III stars may have been born over a longer
period than previously thought, leaving more for us
to find.

Have we therefore reached the limit of stellar archaeology? We've dug down deeper than anyone really expected us to be able to do. There is a rule-of-thumb 'pollution limit' determined by scientist Icko Iben Jr. in 1983[20]. Iben calculated how much metal accretion (metals gravitationally falling on to a star) could occur over a star's lifetime to figure out if these first stars could be right under our noses, just camouflaged, like the dirty red balls in our grabber competition thought experiment. Any measurement of inherent metal lines would always be limited by the amount of 'dirt' that had amassed on a star – even if a metal-poor star existed underneath all the dirt, we couldn't see it. This pollution limit is not definitive and has been broken through already. The Iben calculation was a simplistic one meant only as a guide, but the underlying idea that we will one day reach a limit in what we can detect in terms of metal-poor stars is almost certain, and it doesn't seem too far off.

★ ★ ★

There is real hope that we will one day dig up a surviving Population III star. While the average Population III star is thought to have been tens to hundreds of times the mass of the Sun, simulations show that a tail of low-mass stars was produced at the same time. The more massive a star, the shorter its life, so only stars 80 per cent of the mass of the Sun or less would still be around in the local neighbourhood today. Spotting them isn't easy, but we think that the halo of the Galaxy is a good place to look for them, because there is less contamination there by younger stars, and the older Population III stars will probably have meandered out of the disk and into the halo by now. Generally, iron absorption lines are a good marker of the overall metallicity of a star, and stars with a smaller iron-to-hydrogen fraction than the Sun are called 'metal-poor'.

The progress in stellar archaeology has been significant: scientists have dug down to metallicities of 1/1,000,000th of the Sun's iron abundance, getting to the point where they are not detecting iron at all. The search for metal-poor stars has been challenging, complicated by metal pollution and population mixing, and we are unlikely to make much more progress to lower metallicities. Do not be fooled, though – stellar archaeology is not a dead field, but a field of the dead. It is a field that has moved from seeking the first stars, to conversing with the second stars and hearing tales of their ancestors. As we develop more efficient technology to pick out the most promising candidate stars among the billions of options, the hope is that we will increase our sample size of metal-poor stars and draw more robust conclusions about the diversity of supernovae events that created them. In that way, then, there is the hope that we can reverse engineer a Population III IMF, or at least place constraints on the IMFs coming out of simulation. No matter how hard we search for some artefacts, we just cannot hope to find much more. Thousands of years of wear and tear destroy most fabrics, food and wooden tools. If a population of low-mass stars exists today, it is likely to be polluted, hidden from view, and there is an open question about whether we can ever circumvent that further than we already have.

We may not have detected that very first star, the metal-free star, but we have detected what we think is the first descendent, the star that bridged the populations of metal-free and metal-poor stars. This is not a failure. While we would all love to meet Khufu and ask him how the Great Pyramid was built and where the treasures are buried, I'm sure we'd all settle for a chat with Cleopatra – though saying that, one of my favourite facts about Ancient Egypt is that we are closer in time to Cleopatra than Cleopatra was to the pyramids. The future of stellar archaeology is far from over. It has just found its feet, and with the discovery of SM0313-6708, as we hold our candle up to the skies, we have seen the smallest glint of gold – or should I say iron?

CHAPTER EIGHT

Galactic Cannibalism

A long, long way away, about 75,000 light years away to be exact, lies a small group of stars, called Segue 1. We stumbled upon it only recently, in 2006. Yet in the past decade it has generated quite a lot of controversy due to the claim, rebuttal and renewed claim that this could in fact be the least-evolved galaxy in our Universe. The controversy began not in relation to the question of its age, particularly, but in the 'galaxy' part of the claim. Have we found an ancient fossil, or is this galaxy in fact not a galaxy at all. Is it instead simply a group of stars, younger and less significant?

A good place to try and start to find an answer to this question is to define a galaxy, so let's head to the trusty *Oxford English Dictionary*:

> A feature, system, etc., resembling the Milky Way; (in later use) spec. any of the numerous systems, often of millions or billions of stars, held together by gravitation and containing other matter such as gas and dust, which exist throughout space as distinct bodies.

Now I would argue that this definition is not explicit enough and while 'containing other matter' does include dark matter, it needs to be explicit as it is the dark matter that holds the key to a galaxy's label. This difficulty led to scientists suggesting a new definition of a galaxy in 2012: 'A galaxy is a gravitationally bound collection of stars whose properties cannot be explained by a combination of baryons ['*normal*' *matter*] and Newton's laws of gravity'.[1]

The paper containing the discovery of Segue 1 came out with the title 'Cats and dogs, hair and a hero: a quintet of new Milky Way companions'.[2] Despite announcing a quintet, the title conspicuously names only four objects, namely: Canes Venatici II (Latin for hunting dogs), Coma Berenices (Berenice's Hair), Leo IV (Latin for lion, a cat) and Hercules (the famed hero). According to convention, dwarf galaxies are named after the constellation we observe them in, so Leo IV should be the fourth dwarf galaxy discovered in the Leo constellation. When Segue 1 was first examined, it was so tiny that it was assumed to be a globular cluster – a group of gravitationally bound stars within the Milky Way. Globular clusters are gravitationally bound systems, but nowhere near big enough or massive enough to be called galaxies. They are named after the survey they were found in, so Segue 1 was named thus because it appeared to be the first globular cluster to be found within the Segue survey. Compared with finding four new galaxies close to the Milky Way, a mere globular cluster just didn't make the headline. But Segue 1 wasn't just a group of stars.

Hidden mass

You can measure the mass of a luminous, gravitationally bound object in two ways. The first relates to luminous mass. It involves looking at how much light is coming from a group of stars and figuring out how massive the stars need to be to produce that much light. The second, more complete method, measures the total mass of anything in this star group that might produce a gravitational pull – so all mass, luminous or ... dark. The movement of the stars within a group tells us about this total mass. The physical size of Segue 1 was small, so small that it was easily categorised as a globular cluster. But when a second team of researchers looked into the total amount of mass compared to the amount of light (as a proxy for luminous mass) they found a mass-to-light ratio of 1,340[3], later updated to 3,400[4]. For a globular cluster, the mass-to-light ratio is small, around two or three, differing from one only because of the dead, non-luminous remnants of stars. There is no dark matter bound within globular clusters. Segue 1 went from an uninteresting group of stars that didn't make the title credits, to what we now accept as the most dark-matter-dominated galaxy known. This is very exciting, for two reasons. Firstly, dark matter is a notoriously tricky entity to track down or constrain. We know it is there from kinematic arguments provided by galaxy rotation curves and galactic mass measurements. Finding it in terms of its particle interactions, and therefore defining what it is, has eluded us, however. To have found what appears to be the highest concentration of dark matter that we have ever detected, and right next door to us, has sent particle physicists all abuzz. The second reason that Segue 1's definition as a galaxy is very exciting is that it is small. Very small. And that is interesting because the path to becoming a large galaxy is through galactic cannibalism. Galaxies consume each other, so that over cosmic time we go from a Universe full of lots of little galaxies to one with much larger galaxies. If we've found a small galaxy, then we may have found a survivor, an ancient morsel cast aside by a hungry Milky Way. We've found a first galaxy, and within an aged first galaxy might just lie a lingering first star.

Dwarf galaxies

Segue 1 is not the only dwarf galaxy on our doorstep. The Milky Way is part of a group of galaxies called the Local Group, stretching around 10 million light years across. Galaxies come in lots of different shapes and sizes, so that we can divide them into multiple morphological groups. For example, there are the spirals, the ellipticals, the irregulars and the barred spirals. The Local Group is a 'galaxy zoo', containing a good few of the different types. The three largest galaxies are, in order, Andromeda, the Milky Way and the Triangulum Galaxy, and they are also the only spiral galaxies in the local group. You might have heard of the Large Magellanic Cloud and Small Magellanic Cloud, which are the fourth and sixth most massive galaxies in the Local Group, but aside from them you are unlikely to have come across the others. It's been hard to define when a galaxy is small enough to term it a 'dwarf', just as there isn't an exact definition for when someone is 'short'. We might immediately argue that we could call anyone less than average height short, and anyone greater than average height tall. We certainly cannot apply this to our galaxies, as dwarf galaxies are actually the most numerous galaxies in the Universe – so the nomenclature is a bit unfair. We should really call dwarf galaxies 'galaxies', and the big ones like the Milky

Way 'mega galaxies' or some such, but who am I to argue. Anyway, as a rule of thumb, dwarf galaxies contain a maximum of a few billion stars, compared to the order of several hundred billion stars in the Milky Way, so are much smaller and less luminous.

Just like galaxies, dwarf galaxies divide into a diverse morphology. The Large Magellanic Cloud and Small Magellanic Cloud are both irregular dwarfs, irregular in shape, and not quite fitting into the elliptical, spheroidal or spiral galaxy types. They are young, with plenty of gas fuelling ongoing star formation. Dwarf spheroidals are, in comparison, gas-poor and old, and the ellipticals somewhere in between – elongated spheroids and with some gas left for star formation. In addition, spiral-shaped dwarfs have been observed, but they are thought to be rare. They are easily disrupted through tidal interactions (the gravitational pull felt when orbiting a massive object) with other galaxies, so that the spiral arms are quickly disrupted or dissipated, evolving the galaxy into an irregular or elliptical.* The Large Magellanic Cloud is thought to be an example of this, consisting as it does of only a single spiral arm, after tidal interactions with the Small Magellenic Cloud and the Milky Way.

The interesting dwarfs for us are the dwarf spheroidal galaxies, and the faintest dwarfs of all, ultra-faint dwarfs,

---

* The term for these brief, passing interactions is 'galaxy harassment'. As a campaigner against sexual harassment in academia, this term makes me shudder, but I also must pay kudos to the naming choice. I can just imagine the excuse of the Small Magellenic Cloud: 'I didn't intend to disrupt every part of their being and change them forever, destroying everything that is beautiful and unique, I just have a really large gravitational field. They came towards me!'

UFDs. Dwarf Spheroidals, or dSphs for short, though gas-poor, have a large range in luminosities: from about 20 million down to just a few thousand solar luminosities. In comparison, Andromeda has an estimated luminosity of 10 billion solar luminosities. Observations of dSphs reveal that their supply of gas did not last long enough to form many of the more metal-rich stars that we see forming today. We can peer into these dwarf galaxies and study the stars in the same way that we study stars in the stellar halo in stellar archaeology. The fainter the dSph galaxy we study, the lower the metallicity of the stars within appears to be. Searching for the faintest dwarf galaxies is a focused way of searching for the lowest metallicity stars, and the atmosphere within the field of dwarf galaxy archaeology is one of great optimism and excitement. Stellar archaeology and dwarf galaxy archaeology are complementary methods. Stellar archaeologists dig for discarded artefacts, searching the attic of the Milky Way. In contrast, dwarf galaxy archaeologists search for undisturbed environments. Both provide information of different kinds and work together to complete the puzzle around the first stars. Dwarf galaxies have the potential to be the oldest example of galaxies in our Universe. If we can find one that has been left alone to exhaust its gas and thus forgo any evolution in terms of stellar content, then we can look into a pristine background environment where the Population III stars formed.

Galactic cannibalism
The structures in our Universe formed hierarchically. First the smaller structures, containing just a few stars, merged into larger clusters, or small galaxies. These small

galaxies merged to make larger galaxies, and so on until we see the Milky Way size galaxies. We can see the evidence for this hierarchical galaxy formation, or galactic cannibalism, by observing the galaxy mergers taking place right in front of our noses. There are two examples of ongoing galactic mergers in the centre-fold. The Mice Galaxies and Antennae Galaxies are each examples of two spiral galaxies colliding and ripping apart because of the pull of gravity, producing the long tail of the Mice Galaxies and the antennae of the Antennae Galaxies. When galaxies collide, the stars are all jumbled up inside. Sometimes the new merged galaxy will be bright with new star formation and at other times a loss of gas will cause a quiet period.

When two spirals merge, they tend to be disrupted so much that the spiral arm configuration is lost and the result is an elliptical galaxy. This is probably what will happen to the Milky Way as Andromeda makes its way on a crash course towards us in about 4.5 billion years (incidentally, that's also about when the Sun is due to expire so it's going to be a rubbish time for humanity all round). Dwarfs, then, are the building blocks of the galactic construction site, merging, crashing and accreting on to one another until the resulting structure is no longer a dwarf but a galaxy proper. While the process of galactic cannibalism is necessary to explain the structures we see today, it was not a case of thousands of these dwarfs pinging off each other in every direction. In simulations of hierarchical structure formation for a Milky Way-type galaxy, only three large dwarf galaxies contributed to the formation of the inner regions of the Galaxy. The outer halo regions, where a less violent accretion of passing dwarfs is more likely, had eight lower mass progenitors.[5] We can see evidence of a real–life merger with the Milky Way in a winding structure called the Sagittarius Stream. The Sagittarius Stream results from a long, drawn–out disruption of the Sagittarius dwarf spheroidal galaxy, leaving the stars that once defined it wrapped around the Milky Way in a ribbon. The Milky Way is not kind to the dwarf who wanders too close.

Defining a dwarf

Due to the process of galactic cannibalism, dwarf galaxies are more likely to be old, and have the potential to be leftover building blocks of hierarchical galaxy formation, if we can find them. A lot of dwarfs, though small compared to galaxies like our own, are still obviously galaxies compared to globular clusters. Globular clusters and galaxies were initially easily separable. The former were small and dim, while the latter were brighter and larger. When plotted on a graph of luminosity versus size, the two populations were differentiable. As our telescopes got better and uncovered the ultra-faint dwarfs, however, the two groups overlapped, making it harder to figure out whether a structure was a globular cluster or a dwarf galaxy based only on size and luminosity.

To identify whether a structure is a dwarf, we need to look for two things. First, we measure the velocities of the stars. When orbiting a central mass, the spread in speeds at which the stars travel is proportional to the amount of mass in a galaxy, baryonic and dark. By following the velocities of the stars, we can get an estimate of the mass of a galaxy. We can also estimate the mass of a galaxy by measuring its luminous mass. In this way we achieve two mass estimates: a dynamical, or kinematic, one and a baryonic one. To be a dwarf galaxy, the dynamical mass must be much larger than the baryonic estimate, indicating the presence of dark matter. The difference in mass estimates is the first identifying marker of a dwarf galaxy. The second is a spread in metallicity values for the stars. When metals are expelled from an exploding star, they can be blown out of a small system such as a globular cluster if the combined gravitational force of the system is small enough. For dwarfs, the additional component of the dark matter halo ensures that some or all of the supernova ejecta are pulled back in for use in subsequent generations of stars. This makes those newer generations more diverse in terms of metallicity than their predecessors, and creates a larger spread in metallicity values than seen in globular clusters.

The first galaxies
We may now have the methods to differentiate a small galaxy from a globular cluster, but does this mean that we have found a first galaxy as opposed to just a more modern replica? It is easy to define a first star – it is simply a star that has formed out of primordial gas and is thus metal-free. Defining what a first galaxy is has historically been a little trickier. After all, a binary system of Population III stars in a dark matter halo could make up a galaxy by a loose definition – it's a gravitationally bound stellar system within a halo of dark matter. The problem with calling this scenario a galaxy, though, is

that it is ephemeral. As soon as the stars die and go supernova, then the 'galaxy' is small enough and the dark matter halo paltry enough, that supernova winds soon blow out any gas that was hanging around, and further star formation is not a possibility. To be a galaxy, you also must be able to remain a galaxy for some time. To be a galaxy, you must sustain subsequent generations of star formation,[6] and the first tiny halos with a handful of stars don't qualify.

The cosmological model of the Universe leads us to believe that large structure exists by the accretion and merging of smaller structure, and thus that in the early Universe there was a proliferation of tiny structures. These tiny halos were called minihalos and were around a million times the mass of the Sun. Within the minihalos the first lucky clouds of gas fragmented into the first stars, as described in Chapter 6. In the early Universe, when the rate of mergers was high, the minihalos were drawn to each other gravitationally, merging into larger and larger halos. During this time the first stars were living their brief lives and seeding their environments with metals upon their deaths. When roughly 10 minihalos have merged,[7] we are left with a collection of metal-enriched gas and perhaps some leftover Population III stars, all within a much larger dark matter halo. At this point the minihalos became a much more stable structure called an *atomic cooling halo (ACH)*.

For primordial gas, the only cooling channel available to it was molecular hydrogen. Two hydrogen atoms join to make a molecule, and in the process release energy in the form of a photon that could travel away from the system, cooling the gas overall. This led to fragmentation and the creation of the first stars. With metal-enriched gas, there are more cooling channels available. The mergers of all these minihalos led to a hotter environment, which enabled single hydrogen atoms to become a coolant – and thus these merged stellar systems are known as atomic cooling halos. We consider the ACHs to be the first galaxies in the Universe because they were large enough to withstand the violent environmental effects coming from the deaths of the stars. With the presence of metals, the gas could cool down and fragment on much smaller scales, in the order of the size of our Sun. Within 10–100 million years of a Population III supernova,[8] the hot, metallic gas settled back down and fragmented to form the second generation of stars. Simulations showed that the first set of supernovae seeded their surroundings so efficiently that the second generation was already classed as Population II.[9] It is thought that Population III stars only dominated in galaxies for around 20–200 million years, before Population II became the dominant population only about 400 million years after the Big Bang.[10] It is unlikely, then, that there was a first galaxy consisting entirely of first stars. Instead, the first stars lived in tiny halos, before coming together to form the first dwarf galaxies. These galaxies would swiftly become enriched with metals and a host of Population II stars. They would continue star formation and accreting other dwarfs, becoming larger and larger, to form the gigantic galaxies we see around us today. However, not all dwarfs were swept up in the fray. Roughly 5–15 per cent of the first tiny galaxies in the Galactic neighbourhood are estimated to survive intact as fossil galaxies around the Milky Way.[11]

Dwarf galaxy chemistry

Segue 1 is one of the best candidates we have for a fossil first galaxy. Its size is one pointer, but it is not definitive. The real smoking gun is in the stellar chemistry. Just as they do within our own Galaxy, stellar archaeologists can apply the same techniques to study stars in the dwarf galaxies, to hunt for those metal-poor stars. This is challenging because the dwarfs are so far away compared to the stars we study in our Galaxy. It means that we can pick out only the brightest stars for stellar spectroscopy. In Segue 1, for example, of which only 71 stars were identified as members,[4] only seven are the bright red giant stars that we can perform high-resolution spectroscopy on. Of these, three out of the seven have iron fractions less than 1/1,000th that of the Sun, making them extremely metal-poor stars, and making Segue 1 the least chemically evolved galaxy known.[12] While we have detected stars of greater iron deficiency within our own Milky Way, the search for them has been arduous because they are rare. In Segue 1, an appreciable percentage of the entire stellar system appears to be metal-poor.

The lack of iron is not the only marker of a metal-poor system. All the Segue 1 stars showed enhanced alpha process abundances. The alpha process is one route by which stars convert helium into heavier elements, by successively fusing helium nuclei. A helium nucleus is nicknamed the alpha particle, originating from when Ernest Rutherford, the British nuclear physicist, first differentiated between two types of radioactive particle: alpha (the helium nucleus comprising two protons and two neutrons) and beta (positron or electron). Oxygen, neon, magnesium and silicon are all alpha-process elements, or alpha-elements, because they are all created in this alpha process and their most stable isotopes are multiples of the helium nucleus (i.e. oxygen is four alphas, neon is five alphas). The massive first stars produced alpha-elements in abundance, and upon their deaths returned the metals to the interstellar medium to be used up in the next generation of star formation. As time goes on in a galaxy, the interstellar medium will be enriched with the metals produced within the stars and supernovae. At first, these supernovae are of a specific type: either a core-collapse supernova or pair instability supernova. But eventually, after 100 million years or so, they will have evolved stars of the right make-up to trigger a Type 1a supernova (the explosion of a carbon-oxygen white dwarf). These supernovae produce iron, so after they become a prominent end-state for stars, the iron levels in the galaxy shoot up. As with most things in stellar archaeology, we make our comparisons with iron. The ratio of alpha-elements to iron would be very high for the first stars and their next generation low-metallicity descendents – as the amount of alpha-elements is large but the level of iron is low. Subsequent stars, after Type 1a supernovae contributed iron, will show much lower alpha-iron ratios. So for most galaxies, dwarf or no dwarf, we are used to seeing a plateau in alpha-iron ratio value for the lower metallicity, older stars, then a decline in alpha-iron ratio for the higher metallicity, younger stars. In Segue 1, we do not see this decline at all. This suggests that the later kind of supernovae, the Type 1a supernovae, never contributed significantly to the galaxy ... star-formation was stopped in its tracks.[13] This chemical pattern suggests that Segue 1 really is a fossil galaxy. It is a first galaxy that never quite got going in terms of star formation – that had a few generations of star formation but no more. Segue 1 probably only ever produced 1,500 solar masses of stars over its lifetime – which is miniscule even on dwarf galaxy scales.[14]

Another diagnostic tool that we can use is neutron-capture elements. These heavy elements are formed by neutrons merging with seed nuclei, such as iron. Such neutron-capture elements are thought to form in events that would occur a lot later than the era of the first stars. Neutron star mergers are one proposed production site, for example, whereas in the short lifetimes of the first stars, there hadn't been time for neutron stars to form. Another production site is in the final 1 per cent of the lifetime of an intermediate mass star, when it turns into a so-called asymptotic giant branch (AGB) star. These stars contain inert carbon-oxygen cores surrounded by helium and hydrogen, and will only exist after around a billion years after the first stars, because it takes that long for intermediate-mass stars to live their lives and burn through their fuel. The abundance of neutron-capture elements in a galaxy indicates how many late-time neutron-capture production events occurred.[15] The very low abundance of neutron-capture elements in Segue 1 implies only a single neutron-capture element production event at most,[12] and no fleet of intermediate stars ending their lives in the AGB state, no dance of neutron star mergers expelling heavy elements into the gas. Instead, it seems that the stars within Segue 1 formed out of gas that hadn't been enriched by any of these events ... showing, as is the case with the alpha-elements, that star formation stopped quickly.

The stars that we can observe in Segue 1 will be the lower mass stars that lived long lifetimes. While the stellar content of Segue 1 probably comprised a variety of masses, the higher mass stars have lived their brief lives and their remnants are invisible to us. The current stars represent the low-mass end of the mass range, or initial mass function (IMF), only. Because we have an idea of different possible IMFs, we can infer the previous existence of the high-mass end. When the higher mass stars died, they seeded the gas with metals – between 45 and 900 solar masses worth of metals, depending on which template IMF you use.[12] In contrast, the measured amount of heavy metals in Segue 1 is a mere 0.01 solar masses, so we can conclude that the vast majority of the metallic gas was blown out of the galaxy via the winds of the supernovae that created them. That barely any of these metal expulsions ended up in the surviving Segue 1 stars points to the fact that star formation was shut off really quickly – possibly even by one of the first supernovae to occur. One of the first stars went supernova, expelling the galaxy's gas, leaving Segue 1 with no raw materials to make any new stars. The low-metallicity stars that had already formed were left alone in a tiny galaxy, doomed to remain unchanging for billions of years.

The unique chemical signatures of the dwarf galaxy stars, namely a high alpha-element ratio, a low neutron-capture element abundance and low overall metallicity, provide us with an interesting diagnostic tool. If we find stars that follow a similar chemical pattern within the Milky Way halo, then we might infer that their origin was in a fossil galaxy much like Segue 1. Comparing dwarf galaxy and Milky Way halo stars suggests that up to half of the extremely metal-poor stars in the Galactic halo could have formed in galaxies like Segue 1 with little neutron-capture element abundance. If we take the idea that low neutron-capture abundance can suggest old age, then this marker can be used to search for ancient stars within our catalogues. Ideally, finding stars with zero neutron-capture abundance would indicate one of the very first stars. So far, however, every star surveyed has been found with non-zero neutron-capture element abundance.[16] The search continues.

Dwarf galaxy archaeology
There is a way of identifying the smallest fossil galaxies, and we know that they are some of the least chemically evolved galaxies out there. They seem the perfect target for a search for a metal-free star, a search started with stellar archaeology in our own Galaxy. While the Milky Way overall seems like quite a peaceful place in which to live, there is evidence that it has been through some glancing mergers, disrupting its halo and jumbling up stars from earlier versions of the Milky Way with those contributed by passing dwarfs. This is one reason why looking for metal-free stars in the Milky Way is a challenge: we aren't looking in a pristine environment but in a dusty attic full of assorted and long-forgotten paraphernalia. One set of researchers set about calculating the probabilities of finding a Population III star in various locations.[17] They simulated lots of different

galaxies, of all kinds of masses, from the Milky Way to the tiniest galaxies like Segue 1. In each simulated galaxy, they then counted the number of Population III stars that formed and calculated what percentage of the total number of stars the Population III stars made up, (see the graph labelled 'Finding a first star in dwarf galaxies' in the colour plates section). Each circle on that figure is a simulated galaxy, the colour of which represents the fraction of Population III survivors. The unfilled white circles are simulated galaxies where either no Population III stars formed, or none survived to the present day in the simulation. The vertical lines indicate the mass of each galaxy, corresponding to the top horizontal axis, where mass increases from left to right. What we can see is that the fraction of Population III

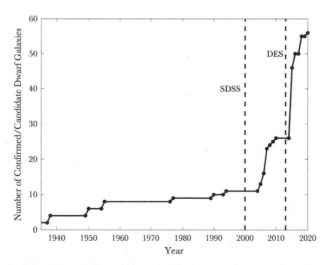

Figure 20 *The number of known dwarfs has recently increased sharply, doubling twice over due to two recent surveys, SDSS and DES. We can expect another doubling shortly, after the launch of LSST. Figure created using data from Simon et al. 2019.*[27]

survivors increases the smaller the mass of the galaxy in question. This is true only up to a certain point, below which there are typically no survivors at all. This figure suggests that, above a certain cut-off mass, dwarfs might be the ideal place to look for metal-free stars, but there's another factor to consider: the observability. In an ideal world we would train our sights on a galaxy and measure the spectrum of every single star, sorting out the metal-free from the metal-rich. There are two problems with this, however. First, we don't have the telescope time. It takes several hours per star to get the spectral information we need, and in galaxies with billions of stars, that's just not an option. Second, in dwarf galaxies, the stars are so far away and faint that we can only detect the brightest of stars with any accuracy. Stars on the lower end of the Population III IMF, at about 0.8 solar masses, would be beginning the end phases of their lives as red giant stars now, and these red giants are luminous enough for us to see, so they are the ones we target. In the figure we can see the Milky Way and its dwarf galaxies, with the filled circles' positions showing the probability of there being at least one red giant Population III star in that galaxy (bottom horizontal axis, where a value of 1 corresponds to a probability of 100%, and a value of 0.5 corresponds to a probability of 50%). For example, the Milky Way has a 100 per cent chance of containing a red giant Population III star, due to its size, while there is less than 50 per cent chance that there is one in the dwarf galaxy Draco. The circles are filled with a colour representing the fraction of Population III survivors. The greater this fraction, the easier it will be to find the star, as there are less irrelevant stars to sort through. You might notice that Segue 1 isn't likely to have any Population III survivors at all – there are just too few stars.

We have a trade-off situation. If we target the smaller dwarfs, there is a greater Population III survivor fraction – but the chances of there being an observable red giant among them are low, because of the small numbers of stars involved. Conversely, if we target the higher mass galaxies, then while we expect there to be more red giant Population III stars, the fraction of Population III survivors is low, and we must sift through a lot of stars to uncover them. The best strategy may lay somewhere in the middle.

Galactic dark matter laboratories

Segue 1 might not be the best place to look for a first star, but it doesn't mean that there isn't value in studying it. Its size was one reason why it was singled out; another was its high dark matter content. Ultra-faint dwarfs like Segue 1 are the closest dark matter halos to us, making them easier targets for dark matter indirect-detection experiments. As introduced in Chapter 5, dark matter is the mysterious component of our Universe, which does not interact with light. The only way we have inferred its presence so far is through its gravitational effects on its surroundings. For example, the presence of dark matter in a galaxy results in the stars moving much faster around the central core than expected. Ideally, we could detect dark matter directly, producing or trapping an atom or molecule of the stuff for further study.

However, because of dark matter's weak or absent interaction with normal matter, we are reduced to finding as much as we can about it indirectly – by inferring its presence gravitationally, for example. To do this, we need laboratories in the sky where we can trial our hypotheses and narrow down the possibilities for what dark matter is and how it behaves. Due to a lack of star formation, feedback effects on UFD dark matter halos were low, keeping the dark matter halos undisturbed and intact. Ultra-faint dwarf galaxies provide the ideal test bed.

Two telescopes have been used in the last few years to stare intently at Segue 1: the Fermi Large Area Telescope (Fermi-LAT) and the Florian Goebel Major Atmospheric Gamma-ray Imaging Cherenkov (MAGIC). The first is a space-based instrument, while the latter is ground based. Both look for the products of Weakly Interacting Massive Particle (WIMP) dark matter annihilation, the proposed process by which two dark matter particles interact and ultimately are destroyed, leaving behind a set of other particles. We don't know for sure what these products will be, but we know that they might produce gamma rays during the process. Gamma rays are the highest energy, or shortest wavelength, type of light. They can travel undeflected by magnetic fields in space. Therefore, unlike most of the possible products produced by dark matter annihilation, these gamma rays have a good chance of making their way to us from a source of dark matter annihilation. If we can detect these gamma rays, then from their energies and number, we can infer a lot about the dark matter and how it annihilates. MAGIC and Fermi-LAT have both looked for gamma rays emanating from Segue 1 hoping to detect something.[18,19] They haven't found anything so far, but as discussed before, a non-detection in itself can tell you something. We call the various ways a dark matter particle can annihilate, channels. For example, there is a tau-anti-tau particle channel and a neutrino-anti-neutrino channel. Particle physicists calculated how much gamma-ray radiation should be detected from each channel for the predicted amount of dark matter in Segue 1. The fact that we detected nothing can narrow down the possibilities and rule out specific channels. What we really want is a detection of a gamma-ray excess over the general background gamma-ray levels of the Universe, revealing that dark matter annihilation is going on. We have detected this kind of excess in one notable place: the centre of the Milky Way.[20] Fermi-LAT detected an excess of gamma rays in the region surrounding the Galactic centre, and this has been widely interpreted as evidence of either an unresolved population of gamma-ray emitting entities – or annihilating dark matter. There is continued observation of multiple dwarf systems because evidence of an excess from any of the dwarfs would cement the view that Fermi-LAT has detected dark matter annihilation in the centre of our own Galaxy. There is also the possibility that annihilation products such as electrons and positrons will produce radio and X-ray radiation. This is trickier to detect as electrons and positrons are charged particles, so are deflected all over the place on their journeys towards us. Still, scientists are trying to detect these by-products, and produced constraints on the WIMPs comparable to those by Fermi-LAT and at complementary low energies.[21,22]

Dwarf galaxies are a great place to test all theories of dark matter, not just WIMPs. MACHOs, or Massive Compact Halo Objects, have been suggested as another kind of dark matter comprising black holes formed in the early Universe.* Black holes are very dark and they make their presence known gravitationally, so it's a valid and interesting idea. MACHOs have the property of injecting energy into a system via its gravitational interaction. If you model a stellar system such as a globular cluster, comprising stars all gravitationally bound together, then add some MACHOs, the increased gravitational interactions cause the orbits of the stars to widen. The system puffs up, sometimes so much that it dissolves into the wider stellar distribution in the galaxy. Because ultra-faint dwarfs are extremely dark matter dominated, you would expect to see this effect clearly. If MACHOs are dark matter, then we would expect to see dwarf galaxy star systems exhibit larger orbits, and conversely, the existence of nice, tightly structured dwarfs, or globular clusters within dwarfs, might disprove the notion of MACHOs. The existence of a nice globular cluster in dwarf Eridanus II (Eri to its friends), and indeed the existence at all of some tinier dwarf galaxies, has ruled out dark matter existing entirely of MACHO particles with a mass greater than about three Moons.[23] That's really tiny for a black hole, and MACHOs are looking less likely than WIMPs.

---

* Yes, physicists did call their two main dark matter contenders MACHOs and WIMPs, and found vaguely justifiable acronyms to do so. The effort that must have gone into that.

The missing dwarfs

Even as the discoveries of the population of dwarf galaxies increased over the past decades, there remained something unsettling about their numbers as a whole. In a cold dark matter Universe, dark matter halos are particularly prolific at the lower end of the mass scale, with hundreds, possibly even thousands,[24] of little dark matter halos predicted to be orbiting the Milky Way. As more sensitive instruments turn on we expect to continue observing more dwarfs, but it's a push to imagine that there are upwards of 1,000 undiscovered dwarfs within the Milky Way halo.

This has historically been called the *Missing Satellite Problem*. While our cosmological understanding of the Universe predicts many hundreds to thousands of dwarf galaxies, we have detected less than 100. Recently, however, it has been proposed that the galaxies aren't missing at all. Just because there is a dark matter halo, this doesn't mean that there is a galaxy. The ionisation of gas by the first stars, and the supernovae winds of the first deaths, provide feedback that can prevent the coalescing of gas into the stars and first galaxies. This feedback will be catastrophic for smaller mass galaxies, and there seems to be a mass cut-off for the halo of about a billion solar masses, below which a galaxy will not form. Researchers corrected simulations for how many halos would remain empty and unseen by current technology. Segue 1 is one of the faintest ultra-faint dwarfs at only 340 times the luminosity of the Sun, and when we assume that all simulated galaxies with luminosity below that limit are unseen, the missing satellite problem goes away. Buzzing around our heads are hundreds if not thousands of dark matter halos, but many of them will be empty, dark clumps of matter that weren't large enough to fight off that initial feedback from the first stars. A larger proportion may even host stars, but the imperceptible luminosity of such faint systems is beyond our observational capacity. In a recent observation, whereas the alternative theory adjusted for this incompleteness bias predicted that two dwarfs should be observed, six popped out of the data, suggesting that the theory still needs tweaking and that the missing satellite problem is now possibly an excess satellite problem.[25]

Small galaxies, cosmic questions

Dwarf galaxy archaeology is a new field, with only a handful of dwarfs known until recently. With a new telescope survey, the Sloan Digital Sky Survey, in the early 2000s, came more discoveries and a doubling of the population of known dwarfs. When the Dark Energy Survey launched in 2013, another doubling followed. The next big telescope relevant for dwarf galaxies is the Large Synoptic Survey Telescope (LSST), which will provisionally start operating in late 2022, or early 2023. This 8m telescope set upon the mountains of Chile will survey the whole sky in only three days, and do that repeatedly over 10 years, creating deeper images – images of farther galaxies. We expect the dwarf galaxy population to explode once again, more than doubling the known current population. A simulations group in Durham, UK, ran cosmological simulations resembling the surroundings of the Milky

Way and Andromeda, and predicted a surfeit of so far undetected dwarfs clustering around the two galaxies in the Local Group.[26] There will be so many dwarfs, all requiring spectroscopic confirmation via measurements of the velocity and metallicity of the stars. This will require high-resolution, time-consuming spectral measurements, and I think to be a dwarf galactic archaeologist will require a good deal of patience in the face of newly discovered treasure.

I have great affection for Durham, as most students do for their respective university towns, and I mirror travel writer Bill Bryson's review of it in *Notes From a Small Island*: 'let me say it now: if you have never been to Durham, go at once. Take my car. It's wonderful'. One of the most beautiful features of Durham is its cathedral, high on a hill above the tiny city. As I sat in my graduation ceremony in that cathedral, then-Chancellor Bill Bryson told us how, over the cathedral's 1,000-year existence, practically every brick had been replaced at some point. The cathedral had evolved and renewed. It strikes me sometimes that the large galaxies we see around us do the same. They may seem to be solid, unchanging, gargantuan entities assembled far in the past, but in fact they are evolving even to this day. The stars we see in these galaxies are not the stars that were there when the Galaxy formed. Perhaps there is the odd ancient star hidden among the more modern replacements, but finding that star is no easy task. Over billions of years, the very structure of the Galaxy may have changed beyond recognition. When we look at the galaxies around us, the incredible diversity we see

speaks to the number of possible histories: it's as if
someone had replaced the bricks in that cathedral with
a random colour and pattern and size. Might it be
possible that one of the earliest galaxies could linger
somewhere in this collection, having avoided all
mergers, having prevented somehow the constant
evolution forced by star formation and circumvented
all catastrophic events that would force it to dissipate?
Segue 1 seems to be the evidence that despite the
violence of the early Universe, some survivors were
preserved among the rubble. With increased telescope
sensitivity, we will discover many more fossil galaxies
and perhaps within a first star.

★ ★ ★

The saying goes that the best things come in small
packages. In the case of dwarf galaxies, the adage certainly
seems to have some truth in it. We have long known of
the existence of dwarf galaxies, those containing a dark
matter halo and a few billion stars at most, but they have
been bright and few in number. In the last two decades
the population of known dwarfs has exploded, doubling
with every new survey. The Milky Way is surrounded by
dwarfs, remnants of the process of hierarchical structure
formation, whereby the larger galaxies were built up
from the merger of smaller structures. The smallest of
dwarfs may be some of the least-evolved galaxies in
the Universe, containing Population III stars. We believe
the first galaxies already contained Population II stars,
having formed from multiple halos of swiftly evolving
Population III stars. The choice to target the smallest of

galaxies in the search for a metal-free star is therefore not clear-cut, as the chances of any of those few stars being in a state to be observable is low. Aside from their astrophysical worth, ultra-faint dwarfs have turned out to be the most dark matter-dominated systems known, and particle physics experiments have turned their sights to an attempt to detect the elusive dark matter particle. With more sensitive experiments due to start observing soon, dwarf galaxy archaeologists will be flooded with data. They will have to carefully choose their targets for spectroscopic follow-up and be patient in the face of holding so many treasure maps, all marked with an X.

In July 2019 I had the privilege of heading back to Durham to attend 'Small Galaxies, Cosmic Questions', a conference on dwarf galaxies. I go to a lot of conferences, but as we tend to stay within our little niches, it is rare to turn up to a conference and know no one. Usually you'll see the same 50 people for the rest of your career. At this conference I was an interloper from the early Universe field, exploring a topic I had only just engaged with, and as a result I knew a total of two people there. Watching a different field and a different set of researchers was fascinating. This is a field that has just begun: there are so many routes of research to explore. At one session it struck me how non-urgent the discourse was. In my field, high redshift 21cm line detection (which is discussed shortly), we are on the verge of a long-awaited detection, all in competition but all willing to be frank and work together to fix the many obstacles in our way. We're in a rush and never have enough time or people to answer all the questions.

At this conference I didn't feel that same urgency or stress. Instead, it had a laid-back feel, a contented excitement, like a child with a pile of opened presents who is unsure which to play with first. I was also struck with how the problems there were presented (jokingly) as learning opportunities. One member of the audience suggested that the mindset should not be to solve these problems, but to create more, to find more things we didn't understand about dwarf galaxies. Until going to this conference, I hadn't appreciated the true breadth of application that the field of dwarf galaxy archaeology has. I discovered that I was not a trespasser in a tight-knit club. Instead, I was encountering a new community comprising multiple families. There were particle physicists who wanted to probe the dwarf galaxies for dark matter. There were stellar archaeologists, excited to have found many more focused locations in which to look for their prized metal-free star. Galaxy formation astronomers were there to use these tiny building blocks to understand how the huge galaxies we see today are formed. And there were other early Universe astronomers like myself, who suggested that younger star-forming dwarf galaxies could be used as analogues of the earliest star-forming galaxies in our Universe. Yet among all this discussion there was little mention of what I thought would be the chief topic: Segue 1. The excitement of finding the first fossil galaxy had waned, and the astronomers were already looking to other candidates and planning more observations with glee. That is not to say that learning about Segue 1 is irrelevant. Quite the opposite. Segue 1 is the first fossil galaxy but it is by no means the last, and the techniques we used to determine its nature are applicable widely. Dwarf galaxy archaeology is a field that is just establishing itself. It is composed of satellite groups of subjects, all orbiting the same goal: to uncover these tiny fossils and use them as laboratories for the understanding of the construction of our Universe.

CHAPTER NINE

The Cosmic Dusk

In a few weeks, I have to cook Christmas dinner for 13 people. It will be a melting pot of immediate family, in-laws, ages, dietary requirements and political beliefs. The way I should fold the napkins and the colour of the tablecloth are suddenly pressing questions, and I have a project-management chart going for preparing the food. It compiles my favourite recipes, telling me when the potatoes need peeling, when the sprouts need steaming, how long my father-in-law's cauliflower cheese will take, when the honeyed carrots (eldest daughter), boiled carrots (the baby), sliced raw carrots (middle daughter) and roast carrots (everyone else) need preparing. Success depends on the perfect execution of each stage. Even with all that expectation there is not too much pressure, as what is the worst that can happen? I burn £50-worth of food and have to break out the emergency mince pies. And yet I feel it, I get stressed about parsnip portions because the responsibility lies with me. I don't understand, then, how people shoulder the responsibility of modern space missions. Take the James Webb Space Telescope,[1] known as JWST. Conceived in 1996, the mission will launch at the earliest in Summer 2021, at a cost of $9 billion.*

---

* As of August 2020. All timings and budgets subject to change.

The telescope has been in planning for such a long time that several of the technologies that it will use did not exist during the formulation stages and were invented for use by JWST. Unlike my culinary worship of the Mary Berry cookbook, those behind JWST are rather more avant-garde in the kitchen, creating delectable dishes on the fly.

The James Webb Space Telescope

The telescope itself reminds me of an old *Transformers* toy you found under the tree on Christmas morning, or perhaps an origami swan napkin on the dinner table. It has an observing surface (called a mirror) 6.5m (21ft) across, which is almost three times as large as Hubble, making it 100 times more powerful. To even fit it on to the rocket they had to come up with a folding mechanism not dissimilar to the leaves on the dining table you extend for festive occasions, whereby the two sides of the mirror will fold neatly away. In space, those mirror leaves have to unpack just so to ensure that the mirror slots together to make a flat surface. Then there is the sun shield, made of five polyimide layers as thin as a human hair but as immense as a tennis court, designed to reflect as much of the heat from the Sun and Earth as possible. The sun shield will be folded 12 times to fit on to the rocket, or rather, a 12-times folded sun shield has to unfold in deep space, exactly as planned, to prevent the experiment being overwhelmed with solar radiation. If you watch a deployment animation of the telescope,[2] you can see how each part of the telescope unfurls, extends, rotates, unpacks and tilts during a whole agonising month following the launch.

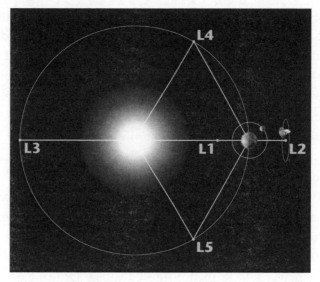

Figure 21 The Lagrange points. Lagrange points are the parking spaces in the Earth-Sun gravitational field. The L2 Lagrange point is of particular use to astronomy because it allows an unimpeded view into space, while remaining close enough to phone home.

JWST will rest at a special point in the Earth–Sun gravitational field called the L2 Lagrange point. There are five special regions in the Earth–Sun gravitational field where the attractive forces of the Earth and Sun precisely balance the orbital motion of a satellite so that it will remain in the same relative position to the two larger bodies. At the L2 Lagrange point, the gravitational pull of both the Sun and Earth precisely balance the outwards repulsive centrifugal force felt by the orbital motion of the satellite. L2 was home to both the Planck telescope and WMAP, the two experiments that have produced beautiful maps of the CMB temperature variations across the sky. It is perfect for space observation because it allows the telescope to shield itself from the Sun and Earth together, while remaining close enough to communicate with Earth. With the Sun and Earth both behind it, the darkness of deep space is revealed. But ... the telescope is awfully far away. With Hubble, when a fault with the mirror became apparent after launch, an admittedly complicated run of repairs carried out by astronauts ensured that it was fixed. Hubble, though, was close, in low Earth orbit. At only 547km (340 miles) above Earth, it wasn't too much trouble to call a mechanic. In comparison, JWST will lie more than 1.5 million kilometres away, four times as far as the Moon, and far, far beyond where any NASA breakdown cover is valid. Due to the need for perfection, each part of JWST has to be tested, retested and checked once more to ensure that everything will go to plan on the big day. This need for perfection has resulted in the budget growing ever larger, from $1 billion at conception to $9 billion as of 2020. As one astrophysicist pointed out, this is only what Americans spend on potato chips annually.[3] Isn't that a small price to pay to watch the earliest objects in the Universe? JWST is high investment, high pay-off. It's cooking Christmas dinner for the Queen, live on television, without a cookbook in sight.

In Chapter 7 we saw how we can use stellar archaeology to build up an idea of how many low-mass Population III stars there were – the stellar initial mass function. The lowest mass stars, about 80 per cent of the Sun's mass, would have lived long enough that they should still be around today: we just have to find them. The higher mass end is more tricky, as those stars will have lived brief lives of only a few million years. That's far short of the 13 billion years they'd need to survive for us to see them in our local neighbourhood. To fill out the larger mass end of the stellar IMF, we have no choice but to look back in time, by looking far away enough that the light we are viewing is that of those first stars, alive, or dying.

One of the most iconic images that Hubble produced is the Ultra Deep Field, an image of which can be found in the centre-fold. Produced in 2004, this incredible snapshot showcases the diversity of 10,000 galaxies in a tiny patch of sky. These 10,000 galaxies all reside in the same area of sky that is obscured if you hold a UK five pence coin, or a US dime, 23m (75ft) away. These galaxies all reside at different distances, and therefore existed as pictured at different times in the Universe's history. The light from the smaller, red galaxies is probably from only a few hundred years after the Big Bang, while the light from the brighter, well-formed spirals is likely to be from closer to a billion years after the Big Bang.

I have a large canvas of the Ultra Deep Field on my living room wall, so I'm very used to looking at it. Every time I do so, my mind struggles to comprehend that the tiny swirls and blobs are all galaxies, frozen in a dance of unimaginable scale. I am awestruck every time. It makes me feel so small. And all that from just a tiny section of sky. I think that JWST will do for age what Hubble did for scale. JWST will pick up the faintest signatures of light from galaxies throughout the Cosmic Dawn, giving us a direct insight into how those galaxies formed and grew. We'll be able to understand, by looking at the

ancestors of the Milky Way, how young we are, as an
individual, as a species, as a planet, even as a Galaxy.
I think we will study an evolutionary timeline of galaxies
and feel the weight of time.

   Despite the succession of JWST as the Next Big
Space Telescope, it isn't a direct replacement for Hubble.
While Hubble is an optical instrument, JWST sees in
infrared. Infrared is a tricky wavelength to play with. The
Earth's atmosphere blocks a lot of infrared wavelengths,
making a James Webb Earth Telescope a dud. You'd want
to get away from Earth anyway, or rather away from all
the Earth-based infrared radiation, which would obscure
the signals we are seeking. So it is that the telescope will
park in space and unfold a giant sun shield to block the
infrared signals from both the Earth and the Sun. With
the Earth and Sun shielded, JWST has a clear, dark view
into the sky, enabling it to detect infrared signatures.
JWST has an impressive roster of science aims. It will
probe the atmospheres of exoplanets, seek the chemical
signatures of life, take snapshots of galaxies throughout
the ages and build up that structure-formation timeline.
In the nearby Universe, it will peer into stellar nurseries,
past all the dust that obscures the eyes of optical
telescopes. It will study star formation as it is happening
right now (well, within the last few thousand years given
the light travel time – near enough). The light from the
early galaxies is redshifted so much that it appears to us
in the infrared, and ideally, we could pick up first stars'
radiation in the same way. Regrettably, even with its
giant folding mirror, JWST does not have the sensitivity
to see individual stars from the Cosmic Dawn. While
there is hope that there might be a few exceptions if

particularly supermassive stars formed,[4] or stars formed in close clusters,[5] so as to produce a brighter combined luminosity, direct exploration of the first stars is unlikely. It is more probable that we will first glimpse the first stars by bearing witness to their deaths.

The fates of our stars

The mass of a star determines how quickly it fuses its fuel, and therefore how long it survives. The mass is also important for determining how a star will die.

Below 8 solar masses: white dwarf For the tiny stars, as the stellar core finishes hydrogen and helium burning, it finds itself at too low a temperature to ignite carbon burning. The core undergoes a collapse and the outer layers are expelled gently into what we call a planetary nebula. The core does not collapse completely and is held up by the pressure of all the electrons in the core whizzing about. This is a white dwarf: a cooling core of a star, held up by the pressure of its electrons. White dwarfs are very dense, as the star collapses substantially before the electron pressure balances the gravitational force. An average white dwarf has the mass of the Sun but squeezed into a space the size of the Earth. The density of a typical white dwarf is about 1,000,000,000 kg/m^3, which roughly equates to your Christmas turkey weighing the same as 3,000 elephants.

Between 8–20 solar masses: neutron star For a more massive star, due to the accompanying increase in internal pressure the stellar core can reach higher temperatures, igniting all nuclear burning stages, for example of carbon, oxygen and silicon. Once silicon burning turns the core to iron, the core collapses, as it cannot fuse iron to create energy – the process uses more energy than it creates. With a greater mass than a white dwarf, the core can continue collapsing even in the face of the pressure generated by the electrons. In the core, the sea of protons and electrons is squeezed to form neutrons, and the pressure they exert is what eventually stops the collapse. The sudden bounce back when the neutrons are formed causes a shock wave to radiate outwards, expelling the outer layers of the star in an explosion called a supernova. This leaves a neutron star, a stellar remnant that will cool and fade. These remnants are even denser than white dwarfs, weighing in at a density of $100,000,000,000,000,000,000 \text{kg/m}^3$. Your Christmas turkey now weighs as much as a small asteroid.

Between 20–100 solar masses: black hole Black holes are formed when the neutron pressure within the core can no longer hold up the outer layers of a star. The collapse continues, compressing the matter so much that it produces an extremely strong gravitational field. Within a certain proximity, called the event horizon, nothing can escape – not dust, not metals, not gas and not even light. The gravitational field is so strong that it would affect your feet more than your head in such a noticeable fashion that you would stretch out like a noodle in a process called *spaghettification*. This strong gravitational field distorts time, making it appear to freeze to an outside observer. Spaghetti-you would appear frozen at the event horizon, falling to your noodly death. Within the event horizon, you would experience time as usual and death would be instant. Your body would disintegrate into its constituent molecules, atoms, electrons, protons and quarks. Don't fall into a black hole.

Black holes aren't the kind of things you can walk up to or poke in a laboratory, and until April 2019 even astrophysicists weren't sure what a black hole looked like. Until that date every image you have ever seen of a black hole had been a simulation or an artist's impression, which I suppose is the same thing, really – it just depends on whether you are using software or wetware to do the simulating. The film *Interstellar* made waves in the scientific community for its realistic depiction of a black hole, with physicists ensuring that the special effects adhered to the physical theories.[6] The term 'black hole' has entered into common use in the English language, despite it describing what feels like a most uncommon entity: a point in space so dense that, within a certain radius, nothing can escape its gravitational pull, not even light. This observation is bounced around a lot: that nothing can escape, 'not even light', so much so that, along with the use of black hole in colloquial language, we have perhaps lost just how incredible this concept is. Light, after all, can put up with an awful lot. Photons have been roaming free since the Big Bang, occasionally interacting with atoms, getting bashed about a bit when they travel through planetary atmospheres, but largely going about their business. It reminds me of one of those fantastic synchronised aerial display teams, beginning in formation, flying at high speeds, and countering turbulence and weather to remain in precise formation over many miles. The Universe doesn't mess with light too much – it lets it get on with things. Light is the hero of this story, after all, as we search for the First Light, as well as it being the principal method by which we observe our Universe, and every hero needs a nemesis, an anti-hero. For us, the black hole provides that role. Light can have been plodding along at a stately 300 million metres per second for 13 billion years, but should it have the misfortune of skirting too close to a black hole, it is suddenly lost forever. Black holes are the bits of our Universe where the standard physics we know and love has given up and gone for a nap, leaving us on the sofa with a kooky, unpredictable aunt in charge, the kind that 'has no rules'.

On Wednesday, 10 April, 2019, I was at a radio astronomy conference close to Jodrell Bank, one of the world's first purpose-built radio astronomy antennas, with 350 other astronomers. We were all sitting in a lecture theatre watching a live screening of the Event Horizon Telescope/EHT[7] press conference announcing the first ever image of a black hole.[8] We watched as the European Research Commissioner stated that this was 'a huge breakthrough for humanity, we are about to take a picture…', and then, no joke, the screen went black. Was this it? After all, black holes are … nope, it was the connection going down. The connection restored and suddenly, there it was. A yellow and orange coffee stain, the Eye of Sauron, the 'gates of hell' as a slightly over-enthusiastic press officer described it. We broke into emotional applause, both because of the scientific wonder of it and because the people who made this image possible were our friends and colleagues. We knew how hard they had worked over many years to ensure that this happened, how many cogs had to fall in to place for such a large-scale experiment to work. The image was created by combining the efforts of eight telescopes distributed around Earth. Two years before, they had all looked in the same direction at the same time, towards the Virgo constellation, to the supermassive black hole within the galaxy M87. Just waiting for all eight locations to have good weather was a significant challenge. These telescopes don't lie around waiting for someone to use them – their time is highly sought after and if the weather isn't good for your session, hard luck. The image of the event horizon of M87's supermassive black hole is an astonishing collaborative feat, and the experiment will go on to produce images of the black hole at the centre of our own Galaxy soon.

Having told you that black holes are points where nothing can escape, how did they image it, then? So far, we have 'seen' black holes using gravity – their gravitational effect on the surrounding Universe. Like spotting someone well known in a shopping arcade, you are unlikely to see that person, but you know that something is happening from the mass of people heading to one point from all directions. We know that there is a black hole at the centre of our Galaxy because the stars in the centre of our Galaxy are all moving as if orbiting a gigantic mass, even though we cannot see anything (yet; the Event Horizon Telescope/EHT will potentially change that). Due to the gravitational pull of the black hole, anything within a certain radius will get drawn towards it, and as it is torn apart by the gravitational forces, the matter forms an accretion disk, just as we saw during star formation. The friction within the disk causes the gas to heat up, radiating so we can see it, and it is that disk that we see, surrounding the shadow of the black hole itself. Black holes are, through their accretion disks, some of the brightest objects in the Universe, but their distance from us makes imaging the silhouette of one a huge accomplishment.

This has been a longer subsection because, quite frankly, black holes are astounding. They transcend any neat historical anecdote or down-to-earth metaphor. Supermassive black holes such as that observed by the EHT are millions or even billions of solar masses. The black holes formed from singular stars are much, much smaller, but their diminutive size does nothing to diminish their wonder. They are the ultimate distortion of the space-time we occupy, and with most stars within the 20–100 solar mass range creating one, there's no shortage of them.

Between 100–260 solar masses: pair-instability supernova In this mass range, the core temperature of a star reaches such high levels after helium burning that it converts some photons into electron-positron pairs in a process called *pair production*. Positrons are the anti-particle equivalent of an electron, which means that an electron and a positron can annihilate into energy, or equivalently out of energy, electron-positron pairs can form. When pair production happens, some photon pressure pushing outwards is lost. The sudden loss of pressure results in a rapid core collapse, as the outer layers rush downwards at the mercy of gravity. For stars below 140 solar masses, there results a pulsing behaviour, as a series of mass shells explodes outwards. For stars above 140 solar masses (and below 260 solar masses), the downwards crush of material is enough that rapid nuclear burning of oxygen and silicon during the core collapse produces an explosion that disrupts the entire being of the star. The whole thing explodes, leaving no remnant whatsoever, no gravestone or even tiny wooden cross. Everything is blown out into the environment, including the metals, making these pair-instability supernovae (PISN) excellent distributers of the metals swept up into Population II. These PISN are some of the biggest thermonuclear explosions found in the Universe. They release about 10,000,000,000 times as much energy as our Sun outputs in its entire lifetime. Despite their brightness, we think they will still be rare in the data because of their faintness over the vast distance: we might detect only a handful over a five-year survey.[9,10]

Above 260 solar masses: direct collapse black holes (DCBHs) These, currently theoretical, entities result from a gas cloud being in the right place at the right time. While most clouds are cooling via molecular hydrogen and forming protostars, there will remain some clouds that haven't quite got to that stage. If such a cloud is situated next to another star-forming site, the radiation from the new stars can flood the cloud, breaking up molecular hydrogen. Within these neighbouring dark halos, molecular hydrogen is destroyed, leaving the cloud unable to cool down, keeping the Jeans mass high and suppressing fragmentation. The whole cloud then collapses as one into an enormous protostar. As accretion is proportional to mass, the cloud accretes up to a solar mass per year from the surrounding gas and quickly collapses into a black hole of mass 10,000–100,000 solar masses. When the black hole forms, it retains about 90 per cent of the Jeans mass that collapsed – pretty much all the cloud ends up as a black hole. I find this idea so intriguing, as we always thought that to get a black hole you needed a star first, but in this scenario the first black holes and first stars formed side by side. It is thought to be a chain reaction as well, as DCBHs produce radiation that dissociates molecular hydrogen in neighbouring halos, causing more DCBHs to form. This goes on for around 150 million years in all,[11] until the stars and DCBHs have pumped out so much radiation that the star-forming gas is decimated and can no longer form either stars or black holes, until the deaths of the first stars produce metals for cooling. Just as in the EHT observation, the accretion disks around DCBHs will signpost their existence so that hopefully JWST will observe them.[12]

Overweight black holes Because of their enormous masses, the first stars are likely to end their lives as pair-instability supernovae or direct collapse black holes. These two phenomena are some of the brightest in the sky, so JWST will have at least a chance of seeing them. It will work in a complementary way to stellar archaeology: by watching the first stars' deaths and categorising how many fall into each mass range, we can fill out the higher mass end of the stellar mass function, inaccessible to stellar archaeology because of the quick lifetimes.

Direct collapse black holes are especially pertinent to study because not only do they give us a way of probing the highest mass end of the stellar IMF, but they also may help us to explain a mystery closer to Earth. Around us we have evidence of supermassive black holes (SMBHs) in lots of galaxies, and looking back in time – as you do – we've even seen SMBHs as far back as only 690 million years after the Big Bang.[13] Except ... how are they already supermassive? You might be familiar with seeing your teenage nieces and nephews on festive occasions and commenting 'haven't you grown up!' Imagine if they turned up at Christmas looking like a 50-year-old. In that weird world, you would have to figure out how they grew so quickly to look middle-aged in only 14 years. There's just not enough time for them to have accumulated that number of wrinkles.

When it comes to black holes, there is a limit to how much mass they can put on.* As more mass is accreted by a black hole, more radiation is emitted by the accretion disk, pushing in-falling material away. At some point the radiative push balances the gravitational pull and the black hole has reached its limit for accretion. We can take the mass of a Population III remnant black hole (10–100 solar masses) and assume that it has been accreting at the maximum limit the whole time (though this is highly unlikely, as it requires a constant supply of gas – gas that is being used up in star formation). Even then, whichever way you calculate it, you cannot make the black hole reach the masses we observe.[14] One way to escape this awkward difficulty is to suggest that SMBHs are the result of DCBHs. DCBHs cheat the system and start at a much higher mass: 10,000–100,000 solar masses. These black holes have time to accrete the gas to form the supermassive black holes we observe,

---

* Sadly there is no limit to the amount of mass I can accrete over the festive period.

providing a neat solution.[15] JWST will shed light on this puzzle once and for all, by allowing us to pin down the prevalence and size of those DCBHs. Like the first stars, DCBHs are a species unto themselves. They can only form over 150 million years or thereabouts, so they, like the first stars, are another lost species that may soon be rediscovered.

Gravitational waves

JWST is not the only telescope aiming to record the deaths of the first stars. Another mega-collaboration of scientists makes up the Laser Interferometer Gravitational-Wave Observatory (LIGO) collaboration.[16] LIGO comprises two experiments in Livingston, Louisiana and Hanford, Washington, US. Each experiment is aiming to detect the gravitational waves, the ripples in space-time, caused by the collision of massive compact objects such as neutron stars and black holes. In Einstein's general relativity, masses emit gravitational waves when they accelerate. The ripples are miniscule, and it is only when two really massive things spiral around each other in the last moments of a collision that the magnitude of the waves is large enough for something like LIGO to pick it up. When a gravitational wave passes through an object, it distorts it. The wave is a distortion in space-time, so you can imagine a wave stretching a ball one way, then stretching it the other way as it passes through. The LIGO experiments are set up to measure this distortion along two perpendicular, identical length arms. Laser light is sent along both arms and reflected off mirrors at the ends of the arms. If a gravitational wave is passing, the lengths of the two arms will differ, and the resulting pattern of light received back will be ever so slightly different. The technology required to do this is phenomenal: the mirrors are held in suspension on wires as thin as a human hair, and even the distant motions of the Moon and Sun cause the mirrors to sway, so magnets are required to keep the mirrors in place.[17]

On 14 September 2015, LIGO detected a signal from a black hole binary system, two black holes orbiting each other, in the last moments before they merged.[18] As they spiralled around each other hundreds of times a second, at a speed approaching that of light, they sent out a characteristic pattern of gravitational waves called a *chirp*. Gravitational waves are so hard to detect that it is only in the last few milliseconds of a black hole's in-spiral that the waves are large enough for us to measure: the change in arm length was the width of a proton. To be sure of a detection that small you need a validating detection. The Hanford and Livingston experiments detected the same signal at different times, but this was only because of their different locations on Earth. Taking that into account, the signals were received at exactly the same time: they backed each other up perfectly. The merger first detected by LIGO was a stellar-stellar one: the black holes had both been created via the 'usual' channel of the collapse of a star of between 20 and 100 solar masses. To detect DCBHs from the Cosmic Dawn, we're going to need a bigger laser. Due for launch in 2034, the Laser Interferometer Space Antenna (LISA) is a space interferometer that will comprise three spacecraft separated in a triangle formation millions of miles wide.[19] The idea is that the frequency of some gravitational wave events is so low that the wavelengths are bigger than Earth itself, so we have to get to bigger distances. As a gravitational wave passes by, the distances between the spacecraft will change minutely. If it all works. Honestly, I don't know how any of these space-based astronomers sleep – I'd be in a state of constant panic.

★ ★ ★

The first stars are likely to have ended their lives as pair-instability supernovae, or they might even have collapsed directly into a black hole early in the protostar formation. We have two upcoming methods of peering into the Cosmic Dawn, or rather the Cosmic Dusk as the first stars set for the last time. The first, a folding infrared telescope hoping to see the supernovae, the second, a space triangle that might shiver as the gravitational waves from colliding DCBH pass by. Both require precision planning and testing, and the funding necessary for that. They are both the types of experiment that will leave astrophysicists sick with nerves as they make their lengthy journeys and unfurl in space. But with that sense of risk come huge rewards. Without the atmosphere of the Earth in the way, they can peer further and into smaller corners of the Universe, uncovering the first stellar deaths, if not the first stars themselves. It'll be like all our Christmases have come at once.

CHAPTER TEN

The Epoch of Reionisation

'O, be some other name! What's in a name? That which we call a rose, by any other word would smell as sweet.' So goes one of the most recognisable excerpts from William's Shakespeare's perhaps most famous play, *Romeo and Juliet*. Juliet is bemoaning the fact that she has fallen in love with a boy with the last name Montague, a family name that is held in contempt by her own family, the Capulets. As she struggles with her feelings, she exclaims that his name is meaningless, and had he been called anything other than Montague, their love would not be forbidden, or 'star-crossed'.

I get Juliet. It's how I feel about the Epoch of Reionisation, the subject of my life's research. The first stars. Hidden black holes. Looking back in time. All of these possibilities sound wonderful. So the fact that they all come under the umbrella of a subject called the 'Epoch of Reionisation' is, well, an anti-climax. What a terrible, terrible name. It took me several months to spell it confidently, let alone pronounce it, and trust me when I say that no one turns up to your public lectures if you advertise it as the topic. Change the title to 'The dark ages of the Universe and the first stars', however, and you're in the money.*

---

* Figuratively. Early on in my career the most I had been paid for a public lecture was a doughnut, and grateful I was for that too.

Reionisation

We have come a long way in our search for the first stars.
We have looked near, using stellar archaeology, and far,
preparing to push even the James Webb Space Telescope
to its limits. As we wait for further success from these
fields, we can continue looking for evidence similar to
the EDGES result. There, the warming of the primordial
hydrogen gas pervading the Universe signalled the
formation of the first stars: the Cosmic Dawn. Now
we apply this idea to slightly later times and use how the
first stars warm and then ionise that hydrogen to deduce
their presence and properties. There are three epochs
that define the construction and lifetimes of the first
stars: the Dark Ages, the Cosmic Dawn and the Epoch of
Reionisation. The Dark Ages is when the gas is just
coalescing to form the first stars; the Cosmic Dawn is
when some of those clouds ignite fusion and become
first stars in their own right; the Epoch of Reionisation
is when the first stars (and black holes and galaxies) heat
and ionise the Universe.

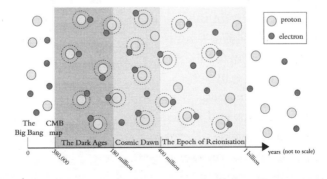

Figure 22 *Over the course of time, the hydrogen pervading the Universe
has changed from ionised to neutral and, thanks to the first stars and
galaxies, back to ionised again.*

After the Big Bang, the Universe was hot. Full of separate electrons and protons and photons all colliding, nothing could settle down. As the Universe expanded and cooled, the electrons fell into company with the drifting protons, forming hydrogen atoms. The photons could travel unimpeded. We say that the Universe became transparent to photons, and formed what we can measure today as the cosmic microwave background. I imagine a train concourse or plaza full of adults (protons) and their hyperactive children (electrons) running about in all directions. As a bystander, it is almost impossible for you (a photon) to walk through to the other side in a straight line, without dodging or diverting. When the adults take control and grab each child by the hand, however, it becomes much easier to walk through without deviating. The Universe was full of primordial gas, predominantly composed of hydrogen atoms, 380,000 years after the Big Bang. When atoms are in a state where the electric charges of the particles (negative for electrons, positive for protons) are balanced, we say the atoms are neutral. The electric charge of the hydrogen atom is balanced, consisting as it does of one positively charged proton and one negatively charged electron. The early Universe was filled with *neutral* hydrogen. Over only a few hundred million years this hydrogen was either wrapped up into the first stars or *ionised*, returned to its constituent state of separate electrons and protons. We call this change in the state of the Universe's hydrogen, from neutral to ionised, the Epoch of Reionisation, returning as it did to the form it was in right in the beginning at the Big Bang. The 're' part of reionisation references the ionised state of the earliest times, just after the Big Bang.

The Epoch of Reionisation was also when the Universe got going in terms of star and galaxy formation. It was the very first stars, galaxies and black holes to exist in our Universe that were responsible for that changing gas. We can flip that on its head and question whether we can use the gas to provide insight on those first objects. The Epoch of Reionisation was a busy time. By the end of this epoch the first stars were dead, the first galaxies were established and the first black holes were guzzling their way to enormity. By revealing the heating of hydrogen by the first stars, the EDGES observation of the Cosmic Dawn provided the warm-up act (pun intended) for a much more complex main event. It's time I stopped hiding behind other titles for fear of losing an audience and present in full glory what this era can do for our understanding of the first light. Welcome, everyone, to the Epoch of Reionisation.

Blowing bubbles

Despite atoms comprising separate parts, namely electrons, protons and possibly neutrons, they are parts that are loath to break apart once formed. To do so requires energy, akin to the energy you would need to pull two really strong magnets apart. Use a small amount of energy and the magnets don't move. Summon more energy, however, and the magnets will separate. It takes different amounts of energy to remove different electrons in different atoms, though for atoms of the same species, like hydrogen, these energy requirements are always the same. For hydrogen, it takes 13.6 electron volts, or 0.000000000000000002 joules, to break away that single electron from its lone proton partner, ionising it. If we refer back to our electromagnetic spectrum, it is only photons with UV wavelengths and shorter that have the equivalent energies to ionise a hydrogen atom. It's a startlingly small energy. I could ionise one hundred billion trillion hydrogen atoms just with the energy inside a Mars bar. It's a big Universe, though, and there are a lot of hydrogen atoms to ionise. These photons have to come from somewhere, and the first stars were efficient photon factories.

At first, there is so much neutral hydrogen surrounding the star that all the UV photons are used up ionising the neutral hydrogen in the immediate vicinity, in a small bubble. As this hydrogen is ionised, the UV photons can travel further before encountering a neutral atom of hydrogen, and in this way gradually larger and larger spherical shells are ionised. Over time, bubbles of ionised hydrogen form around each ionising source. Larger than the stars themselves, and so far more observable, it is these bubbles that are the footprints of the first stars. The

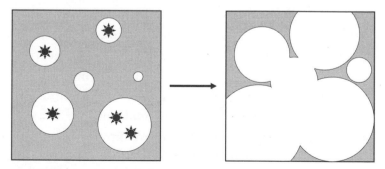

Figure 23 *The Swiss cheese model. Around each first star, a bubble of neutral hydrogen is ionised. Over time, these bubbles become larger and larger, eventually overlapping altogether.*

photons emitted by the first stars tend to stay quite close, ionising hydrogen atoms close to home. This creates nice compact, spherical bubbles. Because of their appearance in the larger swarm of neutral hydrogen, we call this the Swiss cheese model.

Sources of photons: the first stars

The Epoch of Reionisation is a battleground where the toil of photons breaking up hydrogen atoms, *ionisation*, can be swiftly undone by the reforming of hydrogen atoms, *recombination*. Imagine it as being in that room of toddlers again, where half of them are intent on building towers and half are intent on breaking them up. The first stars pump out a lot of UV radiation ... about 10^{50} (that's a 1 followed by 50 zeroes) photons per second,[1] which ionises an awful lot of hydrogen atoms, if you assume that you only need one photon per hydrogen atom to ionise it. But as soon as an atom is ionised, an electron and proton, not necessarily the same ones, can find each other and re-form a neutral hydrogen atom. To maintain a state of reionisation requires a constant supply of new ionising photons. This isn't a problem for the first hundred million years or so of the Epoch of Reionisation, but as the first stars die, the photon factories shut up shop. The Population III stars in their dark matter minihalos didn't constitute galaxies because of their tendency to blow themselves apart with each supernova. When you calculate how many ionising photons these first stars emitted over their lifetimes, it is simply not enough to ionise all the hydrogen in question. Some simulations have suggested that the contribution of Population III stars to reionisation was negligible, while others found that they may have reionised up to 20 per cent of the hydrogen, giving a strong foundation on which to build.[2] However optimistic or pessimistic you may be, upon study of simulations it is clear that we have to seek other ionising sources than just Population III stars.

Sources of photons: quasars

What else was kicking about in the early Universe? Some Population III stars formed black holes upon their deaths. Black holes grow by accreting gas and other normal matter, and as they draw this matter in, the matter forms an accretion disk of migrating matter. This is much like the way a needle tracks on a record, a gentle spiralling inwards. As the matter moves closer to the black hole it is heated because of the strong magnetic fields and friction within the disk, and emits radiation, as anything that is heated is prone to do. We call the system of a black hole and its glowing accretion disk a *quasar*. The accretion disks emit strongly in the X-ray, a more energetic kind of photon than UV. These photons have enough energy to ionise hydrogen atoms too – but unlike the UV photons they also have enough energy to shoot off in random directions and travel a good long distance before they collide with a hydrogen atom. With X-ray reionisation we lose that nice neat Swiss cheese topology. Instead of fully ionised spherical bubbles set in a sea of neutral hydrogen, we get large, partially ionised bubbles and a more uniform and rapid reionisation. In simulations, if we increase the amount of X-rays produced in the early Universe, the result is more wispy bubbles and a faster reionisation overall.[3]

Until recently, it was thought that the quasar population in the early Universe decreased the further back in distance/time we looked.[4] Therefore their contribution to reionisation was thought to be proportionately small. Recent observations of faint quasars at the tail end of reionisation has suggested that there may have been many more quasars in the early Universe than was originally thought.[5] With higher numbers, these quasars could have been responsible for keeping the hydrogen in an ionised state in the end stages, when rare neutral hydrogen islands were holding out against the onslaught of ionising photons. Some have even argued that these quasars were enough to reionise the hydrogen alone, with the contribution from stars forming a negligible proportion.[6]

Moreover, X-rays are not only produced by quasars. Our understanding of the first stars has evolved from seeing them as isolated, single stars, to finding that a significant fraction of the stars may have formed as binaries,[7] or even as more populous systems. In a binary system, the gravitational forces in play can cause one body to accrete matter from the other. When one of these stars accretes matter from its partner, they become what we call an X-ray binary, so named because they emit X-rays during the process. One simulation at least has shown that the X-rays may in fact have had a role similar to or exceeding that of Population III UV emissions,[8] in the very early stages of reionisation. Overall, the contribution of X-ray sources to reionisation remains uncertain. While there has been a recent revival of interest in X-ray sources during reionisation, most simulations point to a third ionising source as the principal character.

Sources of photons: galaxies

As the Population III stars formed and died, and the gas settled, the second generation of stars formed, in larger, merged dark matter halos. These atomic cooling halos were the first galaxies, small in size and hosting a more fragmented, cooler Population II star formation. The Population II stars were higher in metallicity, smaller in mass, greater in longevity and emitted UV photons, albeit at a lesser rate than their ancestors. In the vast majority of simulations, it seems to be the photons emitted from these second generation stars, and together from the galaxy as a whole, which really propelled reionisation, overwhelming the recombination rate and driving the ionised bubbles to grow and overlap, leaving a Universe that was completely reionised. We can gain some insight into this galaxy-driven reionisation by discovering what galaxies were around during reionisation. The Hubble Ultra Deep Field (see image in colour plates section) observed light from galaxies that existed from 800 million years after the Big Bang – right when we believe the Epoch of Reionisation was occurring. By constraining the abundance, luminosities and spectral properties of these early galaxies, Hubble researchers could figure out that this observable population could not reionise the Universe alone. Hubble, like every telescope, has limitations on what it can see. Therefore, it is supposed that there must have been a population of much fainter galaxies, dwarf galaxies, invisible to Hubble, which drove the bulk of reionisation.[9]

Dwarf galaxies are likely to have been the primary drivers of reionisation, but the uncertainty is still there. Even dark matter has popped up again. The annihilation of dark matter (the process by which two particles interact and are destroyed, leaving behind a set of other particles) can theoretically both heat and ionise the gas within dark matter minihalos long before the formation of the first stars.[10] While we think that the overall production of ionising photons is at most on the per cent level,[11] the heating can prevent the gas from fragmenting into stars, delaying star and galaxy formation, and increasing the minimum mass at which stars can form.[12] Dark matter may have had a small effect on how reionisation proceeded, aside from the exotic interactions proposed by EDGES.

There are a lot of different models for how many first stars and black holes there were, how much radiation they output and how long they lived. The reality is that of course all of these phenomena contributed differing amounts at different times. One of the most plausible models for me is that Population III stars pre-ionised the Universe to a small degree before the first galaxies really accelerated the process, as those first stars died. The Universe may have attempted to fight back with recombinations and islands of neutral hydrogen, but those late-forming quasars and their long-reaching X-ray radiation were enough to finish off the job for good. But that's just one theory.

Constraining the Epoch of Reionisation

I have a mouse in my house at the moment. I haven't seen it, but I know it's there. For Christmas I handed my husband his chocolate Santa Claus, only to find that tiny pieces of the wrapping had been torn off, and the chocolate inside gnawed at. The tiny teeth marks were one piece of evidence, but considering that I have three tiny children, it wasn't conclusive. But then, in the bottom of the bag, was more evidence: mouse droppings. Damn, case closed. That we make conclusions or judgements based on evidence is so intuitive that any other possibility seems ridiculous in the present day. There are a lot of distinct possibilities for how these ionised bubbles will look and grow, and we need to narrow down the possibilities based on the evidence. It's a classic whodunnit, or should I say whatdunnit. Something reionised the hydrogen in the Universe – but what? We need more data to constrain our many models.

Constraint number 1: the CMB We can narrow down how long the Epoch of Reionisation lasted using the cosmic microwave background (CMB). The CMB comprises the photons left over from the Big Bang. About 380,000 years after the Big Bang, they could suddenly move freely through the Universe as the temperature cooled enough that neutral hydrogen could form and collisions were rare. As the photons travelled towards us, their energy changed as they fought against the expansion of the Universe, resulting in a well-predicted temperature distribution that Penzias and Wilson, and later missions such as the Cosmic Background Explorer (COBE) mission, Wilkinson Microwave Anisotropy Probe (WMAP) and Planck, could confirm observationally. WMAP and its successor, Planck, could make maps of the CMB over the entire sky, mapping the tiny deviations in the overall temperature. How these deviations appeared, and their magnitude, depends on a number of cosmological factors such as how much dark matter there was, and how much the photons were scattered by free electrons on their travels towards us. This latter factor is of prime importance for us, as the Epoch of Reionisation is nothing if not a gigantic factory of free electrons. We can get a measure for how much the photons have been scattered by looking at the structure in the map.

Consider a photon making its way towards us. If the Universe is full of neutral hydrogen for a long time, then there are no free electrons to collide with until late times. This is a Universe in which reionisation occurs in late times. In a Universe where the reionisation epoch is fast, or very early, there is plenty of time for collisions with free electrons, resulting in a measurable change to the CMB map. We can model how the CMB map would look if different reionisation scenarios had occured and compare to observation. The CMB as observed by Planck matches models where it took only 500 million years for the neutral hydrogen to be completely ionised. Reionisation was rapid.

Constraint number 2: quasar spectra We can narrow down when reionisation ended by looking at the radiation emitted by quasars. Quasars emit radiation across the electromagnetic spectrum. Imagine a quasar was now a small distance away, in our highly ionised Universe. The radiation from the quasar can reach us unimpeded as there are no hydrogen atoms hoovering up photons. We observe the full spectrum that was emitted. Now imagine that the quasar was much further away, and that the light we were seeing was much older and had had to travel through a decidedly different Universe, one filled with neutral hydrogen during or before the Epoch of Reionisation. Neutral hydrogen is an excellent absorber of photons and this creates an absorption line, a little dip in the spectrum, as hydrogen absorbs the photons at a specific wavelength. The further away the quasar, the longer its photons had to battle through neutral hydrogen. As these photons travel, they lose energy, so that what were originally high-energy photons are now at the energy/wavelength that match the energy levels of a hydrogen atom. The hydrogen atoms absorb all the photons of that energy, creating another absorption line. The more neutral the hydrogen encountered between us and the quasar, the more absorption lines appear along the spectrum,

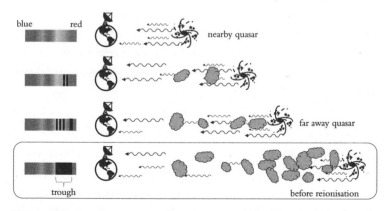

Figure 24 *A quasar emits radiation across a broad range of wavelengths. As the light encounters a neutral hydrogen cloud, the neutral hydrogen absorbs photons, creating an absorption line. The completeness of the resulting trough can indicate the ionisation state of the Universe.*

creating a trough in the spectrum as the absorption lines merge.

When we observe quasar spectra, we generally find that quasars from before or during reionisation have more complete troughs, because the photons have encountered more clouds of neutral hydrogen, so there are more absorption lines. There is a rapid change in the completeness of the troughs, indicating a sudden change in the neutral hydrogen content of the Universe. This change tells us that the Universe rapidly ionised around 1 billion years after the Big Bang. What is also clear is that not all quasars are in the same environment. Some have clean troughs while others observed at similar distances/times have messier ones, indicating that there was less neutral hydrogen at that time in a different place. What this means is that reionisation was patchy. Perhaps unsurprisingly, the first stars and black holes did not act in a synchronised fashion across the gargantuan Universe. In some parts of the Universe reionisation was swift, while in others it was slower. The underlying web pattern spun by dark matter shows us that there were high-density locations where lots of galaxies formed, and low-density voids. It's no surprise that reionisation followed this foundation, with high-density regions ionising first, and low-density regions following after.

We know from CMB data that reionisation was swift, taking only about 500 million years. We know from quasar data that reionisation was finished by 1 billion years after the Big Bang. And we know from Hubble data that large galaxies cannot have reionised the Universe alone. They are some great constraints on the many simulation models we have, all with different ingredients:

Population III stars, X-ray binaries, annihilating dark matter and quasars. Observing the bubbles themselves will provide the most prized constraint and reveal what the properties of all these ionising sources were.

The bubbles of ionised hydrogen are the footprints of the first stars. Sherlock Holmes had the habit of using his trusted magnifying glass to peer more closely at clues, revealing the tiniest detail that would unveil the culprit. There isn't a magnifying glass, well, an optical telescope, large enough to be able to peer back far enough. And even if we had one, it wouldn't be much use when what we are trying to look at is hydrogen. We have mentioned hydrogen a lot so far: it is the most common element in the Universe (not counting whatever dark matter is), and it is the main constituent of stars, fuelling fusion and providing the heat and light. It is quite fitting that the key to detection of the first stars should also be provided by hydrogen. The left-over hydrogen surrounding the first stars gives us an insight into this era with the technology we have now. We may have determined the effect of the first stars on the surrounding hydrogen, but how does that help us to detect it? After all, I could take a photograph of hydrogen on Earth and you wouldn't see anything – it's invisible. But only in the optical wavelengths. Light is a spectrum, and in the radio wavelengths, hydrogen lights up like a Christmas tree.

Radio astronomy
This subject has a bit of a tough reputation compared to other fields of astronomy. Take the Hubble Space Telescope, probably the most famous contemporary astrophysical experiment. It is shiny, and it is in SPACE.

The Hubble Space Telescope is the cool kid at school, the one who effortlessly glides by and whose every word is hung upon. And quite rightly so, really – it's a wonderful feat of human engineering. Most people have heard of the Hubble Space Telescope, and newspapers regularly provided updates of its endeavours. When it briefly broke down in October 2018, it was a top news story and astronomers everywhere were drafted on to TV and radio to reassure viewers that all astronomy wasn't over as a result. In comparison, then, radio astronomy is the quiet child who spends their breaks in an audio-visual room tinkering with old technology for the pleasure of no one but themself. Radio astronomy (at the moment) is a much more down-to-earth pursuit than optical astronomy, consisting largely of metal sticks in fields and deserts, as opposed to shiny clean rooms and rocket launches. But don't make the mistake of writing off radio astronomy as a lesser or obsolete companion to optical astronomy. Radio astronomy, in my opinion, is at the pinnacle of human engineering and scientific pursuit. We have created the infrastructure and computing such that using just some metal sticks in fields and deserts, we can look back and see the footprints of the first stars. Astronomy is time travel, and while optical astronomy is the *Back to the Future* DeLorean, radio astronomy is Dr Who's TARDIS, taking us to times and distant places in our Universe unreachable by any other means.

Radio astronomy is much younger than optical astronomy. Optical astronomy has been around arguably as long as humans have used the stars to navigate and moons to mark the sacrifices. Radio astronomy was born in a much more clear-cut way, out of the Second World War and the use of radar – a technique where radio waves are emitted across a wide area so that any object within that area will reflect back those radio waves. The time it takes for the radio waves to return tells the listener how far away an object is. By producing this wide beam of radio waves, a radar station can detect and track objects within a certain radius, much further away than could be done by eye. The development of radar during the Second World War was a key turning point in favour of Allied forces. Radar stations could provide warnings of bombing raids on London, allowing Londoners to make for shelter as soon as those blips appeared on a screen in a coastal radar station in Kent. Radar was also used as a method to facilitate an attack. Originally, the military had wanted to develop a 'Black Box', a device that would shoot a concentrated beam of radio waves at an enemy plane and simply blast it out of the sky. This was proved unfeasible,* and attention was turned to developing a radar system small enough to sit in a fighter plane. Such radar units could help pilots navigate difficult landscapes and aid them in attacking their targets precisely. They could also help them to seek enemy aircraft within a certain radius, so that reconnaissance planes could sound an early warning, and in an air fight, radar warning screens could warn pilots of an enemy below them or on their tail. Planes fitted with radar units all but destroyed the threat from German U-boats in the space of six short months in 1943. Allied

---

* A £1,000 reward from the Air Ministry for anyone who could kill a sheep at a distance of 100 yards went unclaimed.

planes would locate a surfaced U-boat in complete darkness, only turning on the lights to aid targeting in the last few seconds.

The development of portable radar units was due in no small part to a young physicist called Bernard Lovell. He was drafted into the war effort quite literally overnight, whisked away from his Manchester laboratory in 1939 and planted in various sites in the UK with orders to develop radar units that would be more portable. Lovell worked with colleagues for years on the units, so much so that by February 1945 he was diagnosed with nervous exhaustion and given four weeks rest time, his first proper rest in the war's duration. The view of his incredible dedication and ability was evident in the reply to his request to return to active duty: Albert Percival Rowe, British physicist and one of the higher-ups in the wartime radar effort, wrote: 'You need have nothing whatever on your conscience; if you never did a stroke of work for the rest of your life, you would have justified your existence',[13] which I find both touching and a bit of an odd thing to say.

There is a saying often quoted in science, which is 'One person's noise is another person's treasure.' Lovell spent the war improving the radar units' compactness and ability to detect enemy fighters. The system wasn't perfect and there was an annoying signal interfering with the radar systems, a noise that was thought to come from the radio waves interacting with the ionised part of the upper atmosphere, the ionosphere, creating showers of particles. Lovell had spent his pre-war career trying to investigate properties of particle physics by investigating such particle showers in the lab, using a piece of equipment called a cloud chamber. What Lovell saw with radar was an opportunity to use the sky as a gigantic cloud chamber, and to use radar systems to detect showers of particles in the sky. When the war ended, he started work. Luckily for him there was a surfeit of wartime radio technology lying around, so he could quickly build a simple antenna in a courtyard of the university. Just as quickly, he realised that the built-up confines of the city were far too noisy a place to listen to the quiet Universe. He found some disused land south of Manchester, and initial radio tests showed that this was the perfect radio-quiet environment in which to search for cosmic rays. Once again, he set up his radio equipment and listened. Lovell heard a cacophony, this time not from a terrestrial source but from space. In fact, he heard too much noise for it to fit with the theoretical rate of particle showers: he heard 10 or 12 pings per hour as opposed to the few per night expected. It turned out that Lovell was detecting meteors. A whole new field was born. Before, radar had been used to send signals and receive echos. Now, for the first time, we were listening to what the rest of the Universe had to say.

Despite being distracted by the study of meteors, Lovell did not forget his original mission: to detect the ionisation trails of particles incident on the Earth's ionosphere. It was clear that he needed a more sensitive telescope – not simply an antenna in a field, but a gigantic parabolic surface that would focus a wider area of radio waves on to the receiver, allowing us to listen to quieter signals, well below the hubbub of the meteors. He built that telescope, now called the Lovell Telescope, at Jodrell Bank, and it remains standing as a working icon of radio astronomy. It is home to a fantastic science centre and even an annual music and science festival called Blue Dot (after Carl Sagan's description of Earth as the pale blue dot).

We're going to need a bigger dish

The UK has a soft spot in its heart for radio astronomy, not only because we are part of its origin story but for another well-motivated reason: the weather. Anyone lucky enough to have graced our beautiful islands will know that our weather can leave much to be desired, with rain and cloud a common occurrence. Most international colleagues working in the UK find themselves bewildered and alone in their offices when the weather does reach above 25 °C, as the born and bred Brits know that this is an unofficial Beach Day. The world's largest optical telescopes are based in Hawaii, Tenerife and Chile – all famous for their sunny weather and unlikely to have expensive observations scuppered by cloud cover. The UK, then, is not an ideal place for optical astronomy in general. But radio telescopes can observe in a much greater variety of weather conditions, making them an ideal tool for the UK.

The Lovell Telescope represents the initial stage of purpose-built radio astronomy and the establishment of a sound general principle: space is loud and to hear the quiet stuff you need a big telescope. To listen to a signal from 13 billion years ago, then, a tiny signal buried under the symphony of sounds from every other galaxy in the Universe, we will need a very big dish. Have a quick glance at the images of the 90-m (300-ft) telescope at Green Bank on 15 November 1988 (left) and 16

November 1988 (right) in the colour insert. See the problem? There are engineering limits to the size of these wonderful structures. The Green Bank Telescope functioned for years, producing superb science. But ultimately it was felled by a structural issue – and the larger the telescope, the more prone it is to structural issues. Does this mean that there is a fundamental limit to how small a signal we can listen for, dictated by our ability to build a large enough dish? Luckily for us and the search for the first stars, the answer is no.

Interferometry

When you play a song on your phone, you can do one of two things. You can listen through a speaker or you can listen through some earphones.

Just like listening with two earphones at the same time, it is possible to listen to two antennas* at the same time. The combination of signals from multiple antennas is called *interferometry*. Interferometry solves our problems when it comes to looking back so far in time. By placing antennas 10m (33ft) apart, and combining the signals, we build the equivalent of a 10m dish, roughly speaking. This can scale up pretty much indefinitely, as long as you have the infrastructure to support the connection of the antennas. There is a range of interferometers currently searching the skies for the bubble footprints of the first stars. The Low Frequency Array (LOFAR)[14] is the one closest to my heart, as it was the first telescope I was a part of. LOFAR lies mainly in the Netherlands, where a central island contains a large concentration of antennas,

---

* Always antennas, never antennae. The latter belong to an insect.

and a lower density of antennas stretch out across Western Europe in Germany, the UK, Sweden, Ireland and France. The furthest separation of two LOFAR stations so far is 1,500km (932 miles). Imagine building a single dish of that size.

I visited a LOFAR station for the first time in the summer of 2019, nine years after I began working with it, which about sums up the remote nature of radio astronomy these days. Gone are the days of sitting in cabins and sheds, nervously watching the telescope creak into position. Now we are closer to the space-based optical telescopes, which are designed to work remotely by necessity. It was some time on the seventh leg of my public-transit journey to the home of LOFAR (home–Ebbsfleet–Brussels–Amsterdam–Almere–Kampen Zuid–Meppel–Dwingeloo–Exloo) that I came to the conclusion that it might have been easier to journey into space. The tiles you can see behind me in the centre-fold image are quite underwhelming, aren't they? On the grand scale of things, the tiny LOFAR antenna doesn't amount to much, but the beauty of an interferometer isn't in the individual parts, but in how those individual parts sum up to so much more.

LOFAR consists of almost 2,500 antennas. Through clever combinations of all of those thousands of signals, we can listen in on a tiny signal and observe right back into the Epoch of Reionisation. That was the plan, anyway. LOFAR has been switched on since 2011, yet so far that first stars' signal eludes us. Where is it?

The statistical signal from the stars

Let's consider what we are trying to detect. Ideally, we would be able to look at a patch of sky with enough sensitivity to enable us to get maps of the bubbles, pictures of the Swiss cheese. We could study the topology and pinpoint which of our many simulations with their many ingredients is the 'truth'. With current technology, this is not possible as we don't have enough resolution (see next chapter for the next steps, however), and instead we are looking for what we call a statistical signal, as opposed to a picture.

The neutral hydrogen emits photons of 21cm in wavelength, but that wavelength is stretched as the photons battle against the expansion of the Universe and lose energy. A photon released 400 million years after the Big Bang will be stretched to a wavelength of 2.5m. A photon released a bit later, 700 million years after the Big Bang, will be stretched, or redshifted, to 1.8m. We can tune our radio telescope to pick up only one wavelength at a time, taking samples of light emitted at different times. If we take a sample 400 million years after the Big

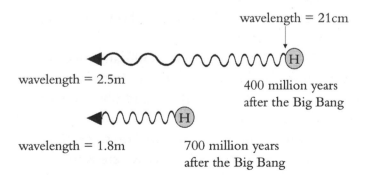

Figure 25 *21cm photons are redshifted to different wavelengths according to when they were emitted. We can tune our radio telescopes to study a particular wavelength and therefore peer into a specific time in the Universe's history.*

Bang, the first galaxies are only just forming and pumping out enough photons to stimulate reionisation. That means that there's lots of neutral hydrogen around, and we can measure a lot of redshifted 21cm radiation. Later on, though, as the Universe is ionised, the photons arriving to us at wavelength 1.8m are rarer as there is less neutral hydrogen and so less 21cm radiation. This is what we seek to detect with current telescopes such as LOFAR: a signal of a lot of 21cm radiation, and then ... nothing. A wiggle and a flatline. Simple, yet we haven't found it, not LOFAR, not the Murchison Widefield Array (MWA) in Australia, not the Precision Array for Probing the Epoch of Reionization (PAPER) in South Africa, not the Giant Metrewave Radio Telescope (GMRT) in India.

The cacophonous Universe
The problem comes down to the Universe being excessively noisy. It is not just the hydrogen surrounding the first stars that emits electromagnetic radiation. There are processes going on in every galaxy in the Universe that produce electromagnetic radiation. When we tune in our radio telescope down here on Earth, we aren't just tuning in to Radio First Stars. We are tuning in to a specific wavelength, and any radiation that arrives to us at that same wavelength all gets lumped together, burying the signal. And there is a lot of it. It's like listening to a phone conversation while walking past a pneumatic drill. The electrons flying around inside our Galaxy are constantly being accelerated and decelerated as magnetic fields affect them. When they decelerate, they produce radio waves, a specific radiation called synchrotron

radiation. This radiation alone swamps our first stars' signals by a factor of several thousand. Then there are the radio signals from every other galaxy. It is a demanding job digging out our tiny signal from underneath all that. We call these astronomical sources of noise *foregrounds*. To make it even worse, the instrument itself is noisy. Electrical equipment produces electromagnetic radiation, and as much as we can protect against it, we can't omit it altogether. The noise from the instrumentation itself swamps our signal by a factor of ten. Consider, if you will, how the townspeople of Gotham City request help from their protector Batman by shining a bat-shaped light into the sky. Luckily for Gotham City, most apocalyptic crime occurs at night, so Bruce Wayne can easily see his signal and stop whatever scheme is underway. But if the Joker starts a scheme in the middle of the day when the Sun is in the sky, then the bat signal is washed out completely, the signal-to-noise is tiny, and the Joker succeeds. It does make you wonder why Bruce Wayne doesn't just set up a phone line.

Foreground mitigation

While the first stars' signal may be much smaller than the signal from the rest of the Universe, the shape of it is different, which gives us a chance to separate it out. Imagine tuning your radio to some soothing classical music, but to your dismay finding a competing pirate radio station interfering on the same frequency. If those pirates are also into Bach and Beethoven, it will be tricky to figure out which of the signals is the legitimate one and which is the pirate radio. But if the pirates instead have a penchant for death metal, then even if their signal is much louder than the legitimate station, you can still fairly easily differentiate the two stations. We use this same idea to separate out the first stars' signal from the sounds of the rest of the Universe, and we do this by looking at how those signals change as we change the wavelength of our observations. If you are used to tuning an FM radio, you will know that a particular station is broadcast on a range of frequencies. BBC Radio 2, for example, is broadcast on 88–91 FM. The cacophony from our own Galaxy and all the other galaxies are similarly broadcast on a range of frequencies (or equivalently wavelength) – a spectrum. If we listen to the synchrotron radiation and slowly change the wavelength, we hear the same sounds but at a different volume – the spectrum is smooth. If we listen to the first stars' signal in the same way, however, we hear distinct sounds as we vary the wavelength. The bubbles of ionised hydrogen grow, merge and recombine, so that every time we take a snapshot the radio emission is different – our spectrum is rough, dissonant. We are listening to classical music at the same time as death metal. The separation of these signals has formed the majority of my research career so far. The methods used to separate them out successfully based on their differently shaped spectra are numerous, and some work better than others in different circumstances, but the fact is that they do indeed work. Despite the overwhelming nature of the problem, the sheer magnitude of the foregrounds compared to our tiny signal, we can remove the foregrounds – but there's more work to be done.

When we apply these techniques, we have to be careful not to declare a signal detection when in fact it is just noise or a residual element of the foregrounds. This is where we must follow evidence-based methodology and account for every piece of evidence thoroughly. This is what my day job is, designing and applying the forensic tests that will remove the foregrounds and one day allow us to be confident about a detection. We have a lot of data by now, petabytes* sitting on a hard disk waiting for someone to search through and find that first signal. But it remains sitting there because our methodology is not yet watertight. We take a small sample of the data and try again and again to reduce those foregrounds, evade that noise, to the best of our abilities. We note how our methods stumble and improve them, starting the whole process again.

The Universe always surprises us

When I meet one of the Principal Investigators (PIs) of the LOFAR Epoch of Reionisation science project, Professor Léon Koopmans, in his office in Groningen in the Netherlands, he is heavily jet lagged from a trip to Australia. Astronomers in the Epoch of Reionisation field had descended on to a tiny island off the coast of Queensland, Australia to share the latest news on their research and updates on experiments such as MWA, EDGES and of course LOFAR. Some scientific fields (no, I won't say which) are famous for being exceedingly competitive, with sniping atmospheres and confrontational conferences, but mine … is just lovely. Despite most of us being in some telescope team or another, the competition is the healthiest it can be – not that criticisms

---

* A petabyte is equivalent to about 2000 years worth of MP3 songs.

and questions are not aired, but they are often followed
by an offer of help or a workable solution. We're a small
field but we all get on rather well, which is a good thing
really considering we will probably find ourselves in the
same room at least twice a year for the rest of our
working lives. We meet in Koopmans' office, one of the
last remaining in the building to have a blackboard. It
is covered in equations, as any self-respecting physicist
ensures at all times. He also has a clock that I immediately
note has only 11 numerals, 55 minutes on the face. He
laughs and says it is because, as an academic, 'you always
have the feeling that you are running out of time', then
fittingly tells me that it took him two weeks to look at
it long enough to notice that it was anything other
than an ordinary clock. Academia is a busy business
and Koopmans is busier than most, co-PI of LOFAR-
EOR, on the management team of the next-generation
radio telescope, the Square Kilometre Array, and involved
in the Netherlands-China Low Frequency Explorer
(NCLE) Moon instrument to boot (the latter two of
which we will look at in the next chapter). Koopmans
has been in the Epoch of Reionisation field since the
conception of LOFAR as an idea, and he recalls the
conversation when they decided to have a go at seeking
this 21cm signal with radio telescopes. His PhD
supervisor-turned-colleague Ger de Bruyn was enthused,
wanting to begin straightaway on the Westerbork
telescopes built in the 1970s, but Koopmans, as he puts
it, has always been a little more cautious. This combination
of enthusiasm and caution was a perfect driver for the
LOFAR-Epoch of Reionisation project that took it to
the first detection. Okay, I'm kidding, we're not there

yet, but I was hoping we would be by the time this book was published, so I defend my use of that as a placeholder sentence. Koopmans's background was gravitational strong lensing, and I asked him why he changed focus to something totally different: 'Well, I like a challenge, and this is a challenge!' The challenges faced by LOFAR, by any Epoch of Reionisation interferometer, have been monumental, and somewhat unexpected. There was optimism as LOFAR was completed in 2012. In 2013 I even published the first Epoch of Reionisation data in my thesis – predominantly noise, but data all the same. More than seven years on, however, and a detection has not been forthcoming. We are the strangest of experiments in that when we say we are searching for a signal, we have in fact already got it. The signal is there, buried in our thousands of hours of observed data, and it has been up to us to brush off the dirt, layer by layer, making sure that we understand each grain that comes away lest we brush off some of our delicate signal by accident. I ask Koopmans if he is confident that we will make a detection with the current-generation telescopes, and he pauses. 'There's certainly excitement, there is still excitement, but I think we are all now much more firmly with our feet on the ground … I think what we are learning is how to do it … My first aim is just to make a detection, that's number 1, and number 2 is … the Universe always surprises us. The data will tell us what the Universe is doing, and I will not be surprised if it is different from the various models.'

Talking of surprises, I ask him about the elephant in every Epoch of Reionisation researcher's room, EDGES. Does he believe it? 'I always say that I am 90 per cent

sceptical and 10 per cent hopeful ... but that 10 per
cent was enough to start with a follow-up and looking
at the most recent results [*from EDGES*] ... it looks like
there is something there.' The enormity of the task
ahead, to uncover that tiny signal, makes scepticism a
healthy viewpoint to have in this field, whether for the
Dark Ages or for the Epoch of Reionisation. Just as
with EDGES, any detection of the Epoch of Reionisation
will need to be checked and validated, probably by
another experiment entirely, before we can claim true
confidence.

★ ★ ★

The Epoch of Reionisation was the period where the
neutral hydrogen pervading the Universe was ionised
by the first stars, galaxies and black holes. Measuring
the degree of ionisation of the Universe at different
times gives us insight into what objects existed at that
time, and how they behaved. The production of
ionising photons was probably dominated by faint
(dwarf) galaxies, with small contributions from Population III stars and quasars. The aim is to observe this
time, and it is the neutral hydrogen that gives us a
window on to it. Neutral hydrogen produces photons
of 21cm in wavelength, but that wavelength is stretched
into radio wavelengths on the photon's journey to us,
so that we must use radio telescopes to observe them.
To peer back 13 billion years we require huge
telescopes, which we build by joining lots of antennas
together across continents in a method called interferometry. There are several telescopes online now, but

the Universe is excessively noisy and it will take time to dig out the tiny first stars' signals from beneath the cacophony.

When will we be done with cleaning the LOFAR data? Good question. In 2010 I thought we would have the detection by 2013. Now, in 2020? I'd put money on it being in the next two years. I'd best get to work.

CHAPTER ELEVEN

Unknown Unknowns

In science there are the known knowns (those that you already know), the known unknowns (those that you are aware of), and the unknown unknowns (those that you didn't even see coming). The data rates of modern telescopes are torrential, and real-time searching of that data for anomalies is near impossible. But if we can store the data and then review it later, sometimes surprises pop out. It is within the archives of these large telescopes that we have made many discoveries, for example *Fast Radio Bursts* (FRBs) within the Parkes telescope.[1] FRBs are exceptionally bright radio bursts, thought to be extragalactic in origin. This field is moving so fast that the favoured model for the origin of FRBs has changed between each draft of this book. There isn't a clear answer yet. The rate of detection of FRBs has increased year on year since their discovery in 2007, as archives around the world have been combed successfully for events, and telescopes commandeered to look for new ones. There is a predicted rate of more than 1,000 FRBs a day across the whole sky.[2] That's a lot, so how on Earth did we miss them? Simply ... because we weren't looking for them. They were an unknown unknown. It was while the archives of the world's radio telescopes were being combed for FRBs that another similar brief, pulse-like event was found. They were named *perytons*, and the data suggested that they were Galactic in origin. This put the FRB community in disarray, as the two types of signal were similar enough for people to question the supposed extragalactic nature of FRBs. Except perytons don't exist. In a beautiful paper, full of good humour and grace, a group of scientists explained that, having looked at perytons closely, 'Subsequent tests revealed that a peryton can be generated at 1.4GHz when a microwave oven door is opened prematurely and the telescope is at an appropriate relative angle. Radio emission escaping from microwave ovens during the magnetron shut-down phase neatly explains all the observed properties of the peryton signals.'[3] Whoops. They hadn't discovered a new astrophysical entity after all. Someone next door had just got hungry while the telescope was taking data. Anything can happen with unknown unknowns!

The known unknowns
We have managed to cover a lot of the known knowns and known unknowns. We have followed the many paths of the first light, whether it came from stars, black holes or galaxies. Simulations have become so advanced that we can now follow the births of individual stars in a fragmenting gas cloud, and follow whole families as they form, merge, fly out or explode. This level of focus was unthinkable a decade ago, and it is this additional level of detail that has led to the shift away from thinking that the first stars were born singularly, doomed to be alone for their brief lives. We now know that they were more likely to have formed in groups, or binaries. The same simulations have led us to realise that low-mass

Population III stars may have formed in addition to their siblings of greater mass – and they might still survive to this day if someone is willing to look for them. Stellar archaeologists have taken on the challenge and continue to set new records for detections of stars with lower and lower metallicities. The field may not have turned up a first star yet, but it has come up with the next best thing: a second generation star. From that alone we can begin to infer the constitution of those original ancestors, and there is no doubt that we will uncover many more among the billions of stars in our Galaxy. Looking further out from our own Galaxy, we have discovered that the multitude of dwarf galaxies in the Local Group could be ideal sites for future extragalactic stellar archaeology once we have the firepower of a telescope able to efficiently survey the faintest of stars in the sky.

As well as searching for the first stars directly, we have employed indirect methods, including performing autopsies on that first generation of stars. You'd think the death of a first star would mean the loss of an observation possibility, but it is quite the opposite. The first stars died in such dramatic ways, either by the super-luminous pair-instability supernovae or by directly collapsing into black holes, that telescopes such as JWST or LISA may potentially detect them. The lives and deaths of the first stars and black holes have a lasting effect on their surrounding environments, an effect that we can measure and reverse engineer to learn about them.

There are three major chapters within the era of the first stars: the Dark Ages, the Cosmic Dawn and the Epoch of Reionisation. The Dark Ages is the time when the Universe was dark and devoid of visible structure. The gas clouds were coalescing, but only when they had collapsed enough to ignite fusion would the Cosmic Dawn unfold, brightening the skies for the first time in hundreds of millions of years. EDGES, a single metal table in the Western Australian desert, took the temperature of the hydrogen 180 million years after the Big Bang and pinpointed the end of the Dark Ages and the beginning of Cosmic Dawn. Not only that, it turned our understanding of gas evolution in the early Universe on its head. Is the famously apathetic dark matter somehow colliding with the gas and cooling it down? Or is there a more terrestrial explanation, a ghost in the machine adding an unwanted signal? Either way, finally the doors have opened on a time previously unexplored.

The Epoch of Reionisation is when those first stars, black holes and galaxies are numerous and powerful enough to ionise the surrounding hydrogen, forming bubbles of ionised hydrogen. This forms a Swiss cheese pattern in the sea of neutral hydrogen, observable to us because of the 21cm radiation that the neutral hydrogen emits. Radio telescopes around the globe have been listening in on the radio waves from that time, but the signals are buried under the cacophony produced by our own Galaxy and other galaxies. After almost a decade of cleaning the data we are close now to a first detection. By observing how these bubbles grow and merge, we can paint a picture of the invisible stars within, stars that died billions of years ago. With current telescopes, we will make what is called a statistical detection: a pinpointing of the time like EDGES but not the imaging of the bubbles that we would ideally seek.

Now what? Ever the question on an astronomer's lips. The sheer scale of a modern-day astronomy experiment requires planning and dedication running back decades, and you can guarantee that as soon as a mission launches, groundwork will begin on the next experiment. As astrophysicists, we are always forced to consider the next step. With the current-generation telescopes such as LOFAR and EDGES, we can only tread so far into the Epoch of Reionisation and the Cosmic Dawn ... and the Dark Ages is almost entirely off limits. For the later stages of the era of the first stars, the Cosmic Dawn and Epoch of Reionisation, we can advance further by using advanced terrestrial technology alone – but for the Dark Ages we are going to need the help of the Moon.

The Square Kilometre Array

Next up in this field is the construction of the Square Kilometre Array,[4] or SKA (pronounced 'Ess-kay-eh' not 'scar' like the music genre). I've been lucky enough to be involved with the SKA through its later design stages and witnessed first-hand the enormous amount of organisation and collaboration that has to occur across the world for years to ensure that in 30 years time we will continue to have new data, new results and new understanding. The Square Kilometre Array is so-named because it is a radio telescope (an array of antennas) with over a square kilometre of observing area, or collecting area, on the sky. It comprises two sets of instruments, one a series of dishes (SKA-MID), and one an array of fixed antennas like LOFAR (SKA-LOW). As well as being split technologically, the experiment is split geographically after a decision was made in 2012 for

South Africa and Australia to co-host it, with the
headquarters in Manchester, England, at the foot of the
Lovell Telescope.

In South Africa, SKA-MID will be constructed within
the Karoo desert region. It will consist of almost 200
dishes up to 150km (93 miles) apart, and will search for
gravitational waves, refine general relativity by searching
for pulsars and even scan the skies for signs of extra-
terrestrial life. The latter science objective is particularly
fascinating to me. Astrochemists will search the regions
around forming planets for spectral signatures of the
building blocks of life, amino acids.[5] It is in these outer
regions of planet formation that these building blocks
could be incorporated into a comet, delivering the
ingredients for life to an inner rocky planet. There is also
a proposal to search for more direct signs of extra-
terrestrial intelligent life by searching for the radio signals
alien civilisations may be emitting. In only 1,000 hours
of observing time, the SKA can scan the nearest 1,000
star systems for signals as quiet as an airport radar on
planets tens of light years away. If someone's there, and
they're emitting at the right frequencies, the SKA should
pick up the signals with ease.

For the Cosmic Dawn, though, it is SKA-LOW in
the Murchison region of the Western Australian desert
that interests us. SKA-LOW will be an aperture array,
which means that it is a large group of connected
fixed antennas, i.e. they can't swivel to point at the
sky. Currently, LOFAR is the largest connected radio
telescope around, with several thousand antennas in the
full experiment. In comparison, the SKA will deploy
131,000 antennas a maximum of 65km (40 miles) apart,

and in a second phase of development this will be closer
to a million antennas. It was one of the central tenets of
the SKA design that the telescope be low on moving
parts because it would be cheap both to deploy and
maintain. I can vouch first-hand for how easy to deploy
these antennas are, as we have one in the shared area of
my office. When it was ordered, it came with instructions
suggesting a swift build time. Clearly, the (lack of)
manual skill of a theory-based astrophysicist wasn't
factored in. It took four of us a couple of hours to
construct it. The antennas are staggeringly cheap too,
only about 200 euros, without the electronics. The final
design for the antenna is not yet confirmed, but the
dominant idea looks like a Christmas tree, so we call it a
Christmas tree antenna (and decorate it accordingly in
December).

Aperture arrays like LOFAR and the SKA can observe
in different directions by combining the antenna signals,
as opposed to steering a dish. Figure 26 shows two
different astrophysical sources giving off radiation. This
radiation is picked up by two antennas, but the signal
arrives at each antenna at different times. The difference
between the two arrival times is called the time delay.
The time delay changes depending on where the source
is in the sky. To point an aperture array in a specific
direction, you select only signals with the time delay
equivalent to the direction you want to observe. In this
way, we say the array can steer its beam (its area of
observation on the sky). With 131,000 antennas, the
operation is extremely complicated and the amount of
data received by the antennas is huge. The SKA is
accelerating technological advances in order to keep up

with the projected data requirements, all while inspiring an endless list of superlative statistics. Two supercomputers, one each based in Australia and South Africa, will archive enough data to fill up 1,000,000 laptops a year. The speed of data flow into the SKA-LOW antennas will be 100,000 times faster than the projected broadband speed in 2022. The optical cables joining up all the antennas to the supercomputers? There's enough of them to wrap around twice the Earth. It's going to be an incredible instrument. Construction should get under way in the next couple of years, and we should see the first scientific observations in 2027. The SKA is the second experiment defining my career, after LOFAR, and I cannot wait until that first torrent of data floods our systems. We are in the enviable position of having too much data, and quick decisions will have to be made about what to keep

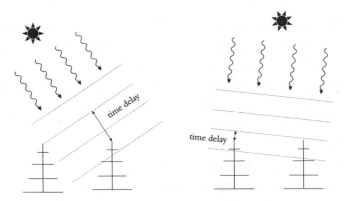

Figure 26 *The time delay between two antennas receiving a signal depends on the source location. This situation can be reverse engineered so that we only analyse signals with the right time delay for the direction we want to look in.*

and what to delete, as there simply isn't enough storage space to keep it all.

With the current generation of telescopes, such as LOFAR and MWA, we aspire to achieve the first detection of the Epoch of Reionisation: a measurement of the temperature of the hydrogen that indicates the formation and eventual overlap of ionised bubbles. These experiments are ultimately limited by the amount of noise that's lumped on top of the signal, noise from the instrument itself and noise from the sky. We can overcome this noise the more observations we take or the more antennas our telescope has. To make an image, you need your signal-to-noise ratio to equal one or more, i.e. the signal must be at least as large as the noise. For the current generation, this is out of reach, so we must be content with making the first ever statistical detections. For the SKA, however, with its 131,000 antennas, the noise is so much lower that the signal-to-noise ratio of one can be achieved pretty much as soon as you turn it on. There'll be some calibration of the telescope to do, but when it is ready for science in the late 2020s, we can expect images of bubbles to appear out of the fog. And the really incredible thing is that we'll be able to take these images over a period of just under 1 billion years, mapping how the Universe evolved. That's an actual film of our Universe growing up over a billion years, right in front of our eyes.

And what will those bubbles look like? Well, it will depend on a lot of variables, both astrophysical (stuff to do with gas) and cosmological (stuff to do with dark matter and the structure of the Universe). We can get help to understand what is inside the bubbles by exploring synergies with other telescopes. For example, if we can study the same section of Universe with the SKA and an optical telescope, then we may relate specific radio bubbles with observed optical galaxies, uncovering the connection between bubble shape and size and what's contained within. The underlying dark matter distribution is also going to affect where the first stars and black holes form. The denser the dark matter, the more gravitational attraction it will have towards the gas and the more likely a star will form. This gives us the tantalising opportunity to use the distribution of bubbles as a way of tracing the dark matter distribution underlying the ancient Universe.[6] That's going to require many, many hours of observations, but the contribution to cosmology will be profound. Astrophysically, the shapes of the bubbles have a lot to do with what source is in the middle. Quasar (black hole) bubbles are more likely to be larger and more irregularly shaped than the neat, spherical bubbles of star-forming galaxies, and most of the bubbles we see are likely to have multiple sources within them. This means that the bubbles don't just contain a single star or a single black hole, but many black holes and/or many stars within many galaxies. We can use the shapes of the bubbles to map the invisible sources within.

With the first phase of the SKA, the 131,000 antennas, we will probe the Universe from about 100 million years after the Big Bang to 1 billion years after it. That's right through the Epoch of Reionisation and starting to get into the Cosmic Dawn. With the second phase of the SKA, which will see upwards of a million antennas, we will make our images even further back into the Dark

Ages. Even with the SKA Phase 1, though, we will be mapping a huge volume of the observable Universe. At the moment our observations are either relatively local, like galaxy surveys, or the single point of observation towards the very beginning, the CMB. As scientists involved in SKA remarked, we are currently making many cosmological conclusions on these data sets assuming 'nothing strange happens in between'[6]. It's quite a thought that our conclusions on, for example, how much dark matter there is in the Universe and how it behaves, are based on such a patchy data set. Incomplete data means incorrect conclusions. The SKA will observe larger volumes of the Universe than have been observed before, making the SKA the perfect cosmological observatory for checking all the models we have on a much larger data set, over a range of volume and times not found elsewhere in astronomy.

To the Moon and back
The SKA is exciting because of its capacity to image the bubbles and its ability to probe right back into the Cosmic Dawn. Its placement in such a sparsely populated, radio-quiet area will make our lives a little easier. Normally we have to spend a lot of effort on removing the human noise of mobile phones, windmills, aircraft, the radio and so on. And we still have to do that to some extent, even in the middle of the desert (microwaves!). This is especially important for future observations of the Dark Ages, the time before the first stars, 100 million years after the Big Bang. To detect light from this far back, we have to tune our radio telescopes down to exceedingly low frequencies, 0.1–30MHz, and this is a

problem for Earth-based telescopes. Above our heads we have layers of atmosphere, the upper layer being known as the ionosphere. The ionosphere is ionised by solar radiation, and the turbulence of the atoms within it seriously affects the transmission of radio waves. For telescopes like LOFAR and the SKA, a large amount of effort is put in to calibrating out the effect of the ionosphere on the observed data, but the problem is an ever-evolving one. The ionosphere changes day by day, hour by hour, because of the variability of the solar wind that ionises it. Below a certain frequency it won't allow the transmission of radio waves at all, and that cut-off can be as high as 10MHz, slap bang in the middle of the Dark Ages window. Radio waves below this frequency just bounce right off of the ionosphere, either back into space, or back down to Earth, depending on where they started. Radio engineers utilise this effect as a neat way of transmitting radio signals across continents by bouncing signals off the atmosphere. This means that the radio frequency interference (RFI) at the Dark Ages frequencies is a nightmare, so much so that it is almost impossible to find a radio-quiet zone anywhere on Earth. And so ... to space!

I can't tell you how exciting the concept of a space mission is to a radio astronomer. We are so used to wryly observing our colleagues' clean rooms and rocket launches – all while getting out our flatpack antennas and readying them for a drive out to the desert on the back of a truck. And boy, have the radio astronomers gone to town on this opportunity. One of the most exciting solutions in play is a radio array on the far side of the Moon. The Moon is a handy object to hide behind because the far side always remains the far side, the 'dark' side, perpetually hidden from Earth. This is because of a phenomenon called tidal locking, whereby the time it takes for the Moon to rotate on its axis matches the time it takes for the Moon to orbit the Earth – so even though the Moon is spinning and the 'dark' side gets just as much sunlight, we only ever see one face of the Moon. The far side is forever hidden from us and, conversely, a radio array placed on the far side of the Moon would be continuously hidden from Earth's cacophony. The US idea, the Farside Array for Radio Science Investigations of the Dark Ages and Exoplanets (FARSIDE),[7] is to have 128 antennas deployed by a rover over 10km (6 miles), while the Lunar Low Frequency Antennas for Radio Astronomy (LUFAR; yes, it's from the people that brought you LOFAR),[8] would ship 50 or more lunar rovers to the Moon, each with an antenna on it. Over a maximum spacing of 10km (6.21 miles) again, the rovers would drive about until they were in the antenna formation required. Imagine looking up at the Moon and knowing that there were 50 little cars doing a dance on the other side. Excitingly, the preliminary steps towards LUFAR have already been taken by a Chinese mission called Chang'e (pronounced 'Chung-ee'), the Chinese Goddess of the Moon. The Chang'e missions are a progressive series of undertakings. Chang'e 1 and 2 achieved lunar orbit, Chang'e 3 achieved a soft landing on the Moon (nothing broke), and it was Chang'e 4 that achieved the first ever fully automated landing on the far side of the Moon on 3 January 2019. The main aim of the mission was to probe the geology of the Moon by landing in one of the deepest and largest craters in the Solar System, so deep that it cuts down into the mantle of the Moon. The geology of the far side of the Moon is thought to be different from that of the near side, and is able to provide us with more clues on how the Moon formed in the earliest stages of our Solar System. For Chang'e 4 to communicate with Earth, it relies on a relay satellite that has a view of both the far side and Earth, at the Earth-Moon L2 point. This relay satellite is called Queqiao, or 'Magpie Bridge', referencing a Chinese folk tale whereby two lovers banished to opposite sides of the Milky Way are briefly reunited every year by a bridge made of magpies. On that relay satellite is an experiment that aims to characterise the radio-frequency interference and general noise background behind the Moon. The Netherlands-China Low Frequency Explorer (NCLE),[9] comprises three 5m (16ft) – long antennas placed on Queqiao. It is taking data as I write, listening very hard to hear what it can on the far side of the Moon, and seeing if it is quiet enough for the proposed experiments. It is the first step on an ambitious road map towards that dance troupe of rovers.

★ ★ ★

Will 21cm radio astronomy make the leap into space in the next few decades? I hope so. Going to space is our only real chance of mapping out the Dark Ages. The Dark Ages are vital for furthering our understanding of astrophysics and cosmology because this was the most pristine time in our Universe. Before the complications of stars, galaxies and planets, there was just a calm

Universe full of dark matter and hydrogen. We can make precise predictions about the state of dark matter, for example, and if our prediction is even slightly off, that points to physics outside our standard models.

The wonderful thing about radio astronomy now is that it combines both the down-to-earth aspects of the past with the most cutting edge research in science. On the one hand, here we are, still with our feet on solid ground, planting metal sticks in remote places. But the data these antennas are gathering is more than our current technology can handle. We are driving tech advances in fibre-optic cables, data storage and cooling mechanisms.

As we constrain our models for the Cosmic Dawn and the Epoch of Reionisation, following decades of simulation and calculation, there's also always the possibility that none of our models will turn out to be the 'right' one. To quote part of the science proposal for the SKA from 2004,

> If history is any example, the excitement of the SKA will not be in the old questions which are answered, but the new questions which will be raised by the new types of observations it alone will permit. The SKA is a tool for as-yet-unborn users, and there is an onus on its designers to allow for the exploration of the unknown.

The SKA has a projected lifetime of about 50 years, so it will take measurements long after I have retired. It is quite a thought that the people who will expertly handle the SKA at the end of its life have not even been born yet. They will be answering questions that we haven't even thought of.

At a talk I gave during the summer of 2019, a secondary school student put up a hand and asked me: 'Does doing cosmology make you depressed?' He was referring to the elements of my talk where I had underlined our infinitesimal size and momentary lifetimes, compared to the ancient and sprawling Universe around us. Sometimes having a job that reminds you daily how insignificant you are can be … wearing. Largely, however, my connection with cosmology has been positive. It makes me feel lucky. I know what a balancing act the path to our existence has been and the serendipity that has helped us to understand the Universe to the level that we do. We've been so resourceful that a piece of rock orbiting our planet has become the doorway to understanding our ancient past, a past we're on the verge of seeing for ourselves. Somehow, these tiny, insignificant organisms have been able to stop fighting among themselves for long enough to plan for an experiment decades away. I feel as though I've been given tickets to the world's greatest spectacle. That's not depressing, that's amazing. Time to enjoy the show.

References

Chapter 1: Over the Rainbow

1 Dvorak, J. 2017. *Mask of the Sun: The Science, History and Forgotten Lore of Eclipses*. Pegasus Books Ltd, Cambridge, UK.
2 www.space.com/37764-new-york-city-1925-total-eclipse.html.
3 Brown, E. W. 1925. The eclipse of January 24, 1925, *Science* 61 (1566): 10–12.
4 Claridge, G. 1937. Coronium. *Journal of the Royal Astronomical Society of Canada* 31 (8): 337–3446.
5 Nordgren, T. 2016. *Sun Moon Earth: The History of Solar Eclipses from Omens of Doom to Einstein and Exoplanets*. Basic Books, New York, US.
6 www.space.com/27412-christopher-columbus-lunar-eclipse.html.
7 Marchant, J. 2009. *Decoding the Heavens: Solving the Mystery of the World's First Computer*. Windmill Books, London, UK.
8 Halley, E. 17145. Observations of the late total eclipse .... *Philosophical Transactions of the Royal Society* 29 (343): 245–262.
9 Brown, E. W. 1926. Discussion of observations of the moon at and near the eclipse of 1925 January 24. *Astronomical Journal*, 37 (866): 9–19.
10 *The New York Times*, vol. LXXIV, no. 24,473, Sunday, 25 January 1925.
11 www.nasa.gov/feature/goddard/2017/chasing-the-total-solar-eclipse-from-nasa-s-wb-57f-jets.
12 Woolf, V., Olivier Bell, A. 1990. *A Moment's Liberty: Shorter Diary of Virginia Woolf*. Chatto & Windus.
13 Klimchuk, J. A. 2006. On Solving the Coronal Heating Problem. *Solar Physics*, 234 (1): 41–77.
14 G Caspi, A. et al. 2020. A New Facility for Airborne Solar Astronomy: NASA's WB-57 at the 2017 Total Solar Eclipse. *The Astrophysical Journal*, 895 (2): id.131.
15 Gleick, J. 2004. *Isaac Newton*. Harper Perennial.
16 Private communication with Royal Society Librarian Keith Moore.

17 Herschel, W. 1800. Experiments on the refrangibility of the invisible rays of the Sun. *Philosophical Transactions of the Royal Society of London* 90: 284–292.
18 Einstein, A. 1905. On a heuristic point of view about the creation and conversion of light. *Annalen Der Physik* 322 (6): 132–148.
19 Gracheva, E. *et al.* 2010. Molecular basis of infrared detection by snakes. *Nature* 464: 1006–1011.
20 Hogg, C. 2011. Arctic reindeer extend their visual range into the ultraviolet. *Journal of Experimental Biology* 214: 2014–2019.
21 Lockyer, N. 1869. Spectroscopic observations of the Sun. No. II. *Philosophical Transactions of the Royal Society of London* 159: 425–444.
22 www.nasa.gov/content/goddard/parker-solar-probe-humanity-s-first-visit-to-a-star.
23 Sobel, D. 1996. *Longitude*. Fourth Estate, London, UK.
24 Rømer, O. 1676. Démonstration touchant le mouvement de la lumière trouvé par M. Roemer de l'Académie des sciences. *Journal des sçavans*: 233–236.
25 Rømer, O. 1677. A demonstration concerning the motion of light, communicated from Paris, in the journal des scavans, and here made English. *Philosophical Transactions of the Royal Society of London*: 893–894.
26 Green, L. 2017. *15 Million Degrees*. Penguin, London, UK.
27 Gamow, G. 1967. *A Star Called the Sun*. Pelican, London, UK.

Chapter 2: Where is Population III?

1 https://abcnews.go.com/Business/beanie-babies-mania-ends-bankruptcy/story?id=19785126.
2 Kahneman, D. *et al.* 1991. Anomalies: the endowment effect, loss aversion, and status quo bias. *Journal of Economic Perspectives*, vol. 5, no. 1: 193–206.
3 Davis, W. 2009. *The Wayfinders: Why Ancient Wisdom Matters in the Modern World*. House of Anansi Press Ltd, Canada.
4 Dempsey, F. 2009. Aboriginal sky lore of the constellation Orion in North America. *Journal of the Royal Astronomical Society of Canada*, vol. 103, no. 2: 65.
5 Sobel, D. 2017. *The Glass Universe*. Harper Collins Publishers, New York, US.
6 Haramundanis, K. 1996. *Cecilia Payne-Gaposchkin: An Autobiography and Other Recollections*. Cambridge University Press, Cambridge, UK.

7   Payne, C. 1924. On the spectra and temperatures of the B stars. *Nature*, vol. 113, 2848: 783–784.
8   Payne, C. 1925. Stellar atmospheres. PhD thesis, Radcliffe College.
9   Payne, C. *The Dyer's Hand*. Privately printed autobiography.
10 Baade, W. 1944. The resolution of Messier 32, NGC 205, and the central region of the Andromeda Nebula. *Astrophysical Journal*, vol. 100: 137.
11 Baade, W. 1944. NGC 147 and NGC 185, two new Members of the Local Group of Galaxies. *Astrophysical Journal*, vol. 100: 147.
12 Russell, H. 1948. On the distribution of absolute magnitude in Populations I and II. *Publications of the Astronomical Society of the Pacific*, vol. 60, no. 354: 202–204.
13 Bond, H. 1981. Where is Population III? *Astrophysical Journal*, v 248: 606–611.

Chapter 3: The Small Bang

1   Watanabe, S. *et al.* 1995. Pigeons' discrimination of paintings by Monet and Picasso. *Journal of the Experimental Analysis of Behavior*, vol. 63: 165–174.
2   Scarf, D. *et al.* 2016. Orthographic processing in pigeons (*Columba livia*). *Proceedings of the National Academy of Sciences Sep 2016*, 201607870.
3   Scarf, D. *et al.* 2011. Pigeons on par with primates in numerical competence. *Science* 334, (6063): 1664.
4   Levenson, R. *et al.* 2015. Pigeons (*Columba livia*) as trainable observers of pathology and radiology breast cancer images. *PLoS ONE* 10(11): e0141357.
5   Armando, the 'Lewis Hamilton of pigeons' sells for record €1.25m. March 2019. www.bbc.co.uk/news/world-europe-47610896.
6   Blechman, A. 2017. *Pigeons: The Fascinating Saga of the World's Most Revered and Reviled*. Grove Press, New York, US.
7   For Heavens sake stop it. May 2010. www.lettersofnote. com/2010/05/for-heavens-sake-stop-it.html.
8   'Mary of Exeter'. www.pdsa.org.uk/get-involved/dm75/ the-relentless/mary-of-exeter.
9   The pigeon, the antenna and me: Robert Wilson. October 2015. www.scientificamerican.com/video/the-pigeon-the-antenna-and-me-robert-wilson.

10 Doyle, A. 2015. *The Sign of Four*. Penguin English Library, London, UK.
11 Chown, M. 1993. *Afterglow of Creation*. Arrow Books, London, UK.
12 The Big Bang's echo. *NPR News*. May 2005. www.npr.org/templates/transcript/transcript.php?storyId=4655517.
13 What is a Cosmological Constant? https://wmap.gsfc.nasa.gov/universe/uni_accel.html.
14 Van der Marel, R. *et al*. 2012. The M31 velocity vector. III. Future Milky Way M31-M33 orbital evolution, merging, and fate of the Sun. *The Astrophysical Journal* 753 (1).
15 Hubble, E. 1929. A relation between distance and radial velocity among extra-galactic nebulae. *Proceedings of the National Academy of Sciences of the United States of America*, vol. 15, issue 3, 168–173.
16 Lemaître, A.G. 1931. A Homogeneous Universe of Constant Mass and Increasing Radius accounting for the Radial Velocity of Extra-galactic Nebulæ. *Monthly Notices of the Royal Astronomical Society*, vol. 91: 483–490.
17 The steady-state challenge. www.britannica.com/science/astronomy/The-steady-state-challenge.
18 Dicke, R. *et al*. 1965. Cosmic black-body radiation. *The Astrophysical Journal*, vol. 142: 414–419.
19 Penzias, A. & Wilson, R. 1965. A measurement of excess antenna temperature at 4080 Mc/s. *The Astrophysical Journal*, vol. 142: 419–421.
20 'The Nobel Prize in Physics 1978'. www.nobelprize.org/prizes/physics/1978/summary.
21 'The Nobel Prize in Physics 2019'. www.nobelprize.org/prizes/physics/2019/summary.
22 McAllister, A. 2016. *A Year Full of Stories: 52 Folk Tales and Legends from Around the World*. Frances Lincoln Children's Books, London, UK.
23 Weinberg, S. 1993. *The First Three Minutes: A Modern View of the Origin of the Universe*. Basic Books, New York, US.

Chapter 4: A Lucky Cloud of Gas

1 Okinawa's annual tug-of-war requires lots of workers, and rope. October 2006. www.stripes.com/news/okinawa-s-annual-tug-of-war-requires-lots-of-workers-rope-1.54989.

2 Rope breaks for the first time at annual great tug-of-war.
   October 2019. http://english.ryukyushimpo.jp/2019/10/
   18/31190.
3 Adams, D. 2016. *The Hitchhiker's Guide to the Galaxy*. Pan, London,
   UK.
4 *The Apollo 15 Hammer-Feather Drop*. July 2018. https://moon.
   nasa.gov/resources/331/the-apollo-15-hammer-feather-drop.
5 Two lose arms in Taiwan tug-of-war. 25 October 1997.
   The Nation.
6 Caroll, B. & Ostlie, D. 2017. *An Introduction to Modern Astrophysics*.
   Cambridge University Press, Cambridge, UK.
7 Grossman, D., Ganz, C. & Russell, P. 2017. *Zeppelin Hindenburg:
   An Illustrated History of LZ-129*. The History Press, Cheltenham,
   UK.
8 Hydrogen and helium in rigid airship operations. www.airships.
   net/helium-hydrogen-airships.
9 Green, L. 2017. *15 Million Degrees*. Penguin, London, UK.
10 Charles 'Don' Albury, 84. *Time Magazine*, 25 July 2005.
11 Phillips, A. 2010. *The Physics of Stars*. Wiley.
12 Fukushima melted fuel removal begins 2021, end state unknown.
   https://abcnews.go.com/International/wireStory/fukushima-
   melted-fuel-removal-begins-2021-end-state-67426592.
13 www.iter.org.
14 Mužić, K. *et al*. 2017. The low-mass content of the massive young
   star cluster RCW 38. *Monthly Notices of the Royal Astronomical
   Society* 471 (3): 3699–3712.

Chapter 5: The Dark Ages

1 http://katiepaterson.org.
2 http://katiepaterson.org/portfolio/the_cosmic_spectrum.
3 Baldry, I. *et al*. 2002. The 2dF galaxy redshift survey: constraints on
   cosmic star formation history from the cosmic spectrum.
   The Astrophysical Journal 569 (2).
4 Falchi, F. *et al*. 2016. The new world atlas of artificial night sky
   brightness. *Science Advances* 2 (6).
5 www.theguardian.com/science/2018/feb/28/cosmic-dawn-
   astronomers-detect-signals-from-first-stars-in-the-universe.
6 https://loco.lab.asu.edu/edges.
7 Bowman, J. *et al*. 2018. An absorption profile centred at 78
   megahertz in the sky-averaged spectrum. *Nature* 555: 67–70.

8 Barkana, R. 2018. Possible interaction between baryons and dark-matter particles revealed by the first stars. *Nature* 555: 71–74.

9 Vogelsberger, M. *et al.* 2014. Introducing the Illustris Project: simulating the coevolution of dark and visible matter in the Universe. *Monthly Notices of the Royal Astronomical Society*, vol. 444, 2: 1518–1547.

10 This is how much dark matter passes through your body every second. July 2018. www.forbes.com/sites/startswithabang/2018/07/03/this-is-how-much-dark-matter-passes-through-your-body-every-second/#7cb9baaf7ccd.

11 http://katiepaterson.org/portfolio/all-the-dead-stars.

12 Sidhu, J. S., Scherrer, R. J. & Starkman, G. Death and serious injury from dark matter. *astro-ph/arXiv*: 1907.06674.

13 Bernabei, R. *et al.* 2018. First model independent results from DAMA/LIBRA-Phase 2. *Nuclear Physics and Atomic Energy*, vol. 19, issue 4: 307–325.

14 Muñoz, J. B. & Loeb, A. 2018. A small amount of mini-charged dark matter could cool the baryons in the early Universe. *Nature* 557: 684–686.

15 Ewall-Wice, A. *et al.* 2018. Modeling the radio background from the first black holes at Cosmic Dawn: implications for the 21cm absorption amplitude. *The Astrophysical Journal* 868: 63.

16 Fixsen, D. J. *et al.* 2011. ARCADE 2 Measurement of the absolute sky brightness at 3-90 GHz. *The Astrophysical Journal*, vol. 734: 11.

17 Dowell, J. & Taylor, G. B. 2018. The radio background below 100 MHz. *The Astrophysical Journal Letters*, vol. 858: 6.

18 http://katiepaterson.org/portfolio/earth-moon-earth.

Chapter 6: Fragmenting Stars

1 www.epa.gov/greatlakes/facts-and-figures-about-great-lakes.

2 Toledo water clears, but outlook is cloudy. August 2014. *The Wall Street Journal*. www.wsj.com/articles/toledo-mayor-orders-more-drinking-water-tests-1407141074.

3 Kopp *et al.* 2005. The Paleoproterozoic snowball Earth: a climate disaster triggered by the evolution of oxygenic photosynthesis. *Proceedings of the National Academy of Sciences of the United States of America* 102 (32): 11131–11136.

4 Schirrmeister, B. *et al.* 2013. Evolution of multicellularity coincided with increased diversification of cyanobacteria and the Great Oxidation Event. *PNAS* January 29, 110 (5): 1791–1796.
5 Jeans, J. 1928. *Astronomy and Cosmogony.* Cambridge University Press, Cambridge, UK.
6 Phillips, A. C. 2010. *The Physics of Stars.* Wiley.
7 Whale explodes in Taiwanese city. Jan 2004. http://news.bbc.co.uk/1/hi/sci/tech/3437455.stm.
8 Tajika E. & Harada M. 2019. Great oxidation event and snowball Earth. In: Yamagishi A., Kakegawa T. & Usui T. (eds), *Astrobiology.* Springer, Singapore.
9 Walker, G. 2014. *Snowball Earth.* Bloomsbury, London, UK.
10 Clark, P. 2011. The formation and fragmentation of disks around primordial protostars. *Science,* vol. 331, issue 6020: 1040–.
11 Loeb, A. 2010. *How Did the First Stars and Galaxies Form?* Princeton University Press, New Jersey, US.
12 Susa, H. 2019. Merge or survive: number of Population III stars per minihalo. *The Astrophysical Journal,* vol. 877, issue 2, article id. 99: 10 pp.
13 Hosokawa, T. *et al.* 2016. Formation of massive primordial stars: intermittent UV feedback with episodic mass accretion. *The Astrophysical Journal,* vol. 824, issue 2, article id. 119: 26 pp.
14 Greif, T. *et al.* 2012. Formation and evolution of primordial protostellar systems. *Monthly Notices of the Royal Astronomical Society,* vol. 424, issue 1: 399–415.
15 Stacy, A. & Bromm, V. 2013. Constraining the statistics of Population III binaries. *Monthly Notices of the Royal Astronomical Society,* vol. 433, issue 2: 1094–1107.
16 Caroll, B. & Ostlie, D. 2017. *An Introduction to Modern Astrophysics.* Cambridge University Press, Cambridge, UK.
18 Muratov, A. *et al.* 2013. Revisiting the first galaxies: the epoch of Population III stars. *The Astrophysical Journal,* vol. 773, issue 1.
17 Wise, J. *et al.* 2012. The birth of a galaxy: primordial metal enrichment and stellar populations. *The Astrophysical Journal,* vol. 745, issue 1, article id. 50: 10 pp.

Other useful references:
Greif, T. 2014. The numerical frontier of the high-redshift Universe. *Computational Astrophysics and Cosmology,* vol. 2, article id.3: 36 pp.

Hallam, T. 2014. *Catastrophes and Lesser Climates, The Causes of Mass Extinctions.* Oxford University Press.

Klessen, R. 2019. *Formation of the first stars,* Formation of the First Black Holes. Ed Latif, M.; Schleicher, D. World Scientific Publishing Co. Pte. Ltd, Singapore. pp. 67–97.

Chapter 7: Stellar Archaeology

1 www.nationalgeographic.com/history/magazine/2018/03-04/findingkingtutstomb.

2 Comelli, D. *et al.* 2016. The meteoritic origin of Tutankhamun's iron dagger blade. *Meteoritics & Planetary Science* 51, no. 7: 1301–1309.

3 Growth reference data for 5–19 years. www.who.int/growthref.

4 Evolution of adult height over time. www.ncdrisc.org/data-downloads-height.html.

5 Stulp, G. 2015. Does natural selection favour taller stature among the tallest people on Earth? *Proceedings of the Royal Society B.* 282, 1806.

6 Habicht, M. E. *et al.* 2015. Body height of mummified pharaohs supports historical suggestions of sibling marriages. *American Journal of Physical Anthropology* 157, 3.

7 Stacy, A. & Bromm, V. 2013. Constraining the statistics of Population III binaries. *Monthly Notices of the Royal Astronomical Society,* vol. 433, issue 2: 1094–1107.

8 Stacy, A. *et al.* 2016. Building up the Population III initial mass function from cosmological initial conditions. *Monthly Notices of the Royal Astronomical Society,* vol. 462, issue 2: 1307–1328.

9 Asplund, M. *et al.* 2009. The chemical composition of the Sun. *Annual Review of Astronomy & Astrophysics,* vol. 47, issue 1: 481–522.

10 Frebel, A. 2015. *Searching for the Oldest Stars: Ancient Relics from the Early Universe.* Princeton University Press, New Jersey, US.

11 Frebel, A. 2018. From nuclei to the cosmos: tracing heavy-element production with the oldest stars. *Annual Review of Nuclear and Particle Science,* vol. 68, issue 1: 237–269.

12 Lawrence Hugh Aller 1913–2003, a biographical memoir. www.nasonline.org/publications/biographical-memoirs/memoir-pdfs/aller-lawrence.pdf.

13 Chamberlain, J. & Aller, L.H. 1951. The atmospheres of A-type subdwarfs and 95 Leonis. *Astrophysical Journal,* vol. 114: 52.

14 Roman, N. 1950. A correlation between the spectroscopic and dynamical characteristics of the Late F- and Early G-Type stars. *Astrophysical Journal*, vol. 112: 554.

15 Baade, W. 1944. The resolution of Messier 32, NGC 205, and the central region of the Andromeda Nebula. *Astrophysical Journal*, vol. 100: 137.

16 Frebel, A. & Norris, J. 2013. *Metal-poor Stars and the Chemical Enrichment of the Universe, Planets, Stars and Stellar Systems 5*, by Oswalt, Terry D.; Gilmore, Gerard. Springer Science+Business Media Dordrecht, Berlin, Germany. p. 55.

17 Lucey, M. *et al.* 2019. The COMBS survey – I. Chemical origins of metal-poor stars in the Galactic bulge. *Monthly Notices of the Royal Astronomical Society*, vol. 488, issue 2: 2283–2300.

18 Nordlander, T. *et al.* 2019. The lowest detected stellar Fe abundance: the halo star SMSS J160540.18-144323.1. *Monthly Notices of the Royal Astronomical Society: Letters*, vol. 488, issue 1: L109–L113.

19 Keller, S. *et al.* 2014. A single low-energy, iron-poor supernova as the source of metals in the star SMSS J031300.36-670839.3. *Nature*, vol. 506, issue 7489: 463–466.

20 Iben, I. 1983. Open questions about the formation of heavy elements in 'Z = O' stars. *Memorie della Societa Astronomica Italiana*, vol. 54: 321–330.

Other useful references:

Christlieb, N. et al. 2002. A stellar relic from the early Milky Way. *Nature*, vol. 419, issue 6910: 904–906.

Frebel, A. & Norris, J. 2015. Near-field cosmology with extremely metal-poor stars. *Annual Review of Astronomy and Astrophysics*, vol. 53: 631–688.

Chapter 8: Galactic Cannibalism

1 Willman, B. & Strader, J. 2012. 'Galaxy,' Defined. *The Astronomical Journal*, vol. 144, issue 3, article id. 76L: 12 pp.

2 Belokurov, V. *et al.* 2007. Cats and dogs, hair and a hero: a quintet of new Milky Way companions. *The Astrophysical Journal*, vol. 654, issue 2: 897–906.

3 Geha, M. *et al.* 2009. The least-luminous Galaxy: spectroscopy of the Milky Way Satellite Segue 1. *The Astrophysical Journal*, vol. 692, issue 2: 1464–1475.

4   Simon, J. *et al.* 2011. A complete spectroscopic survey of the
    Milky Way Satellite Segue 1: the darkest Galaxy. *The Astrophysical
    Journal*, vol. 733, issue 1, article id. 46: 20 pp.

5   Fattahi, A. *et al.* 2020. A tale of two populations: surviving and
    destroyed dwarf galaxies and the build up of the Milky Way's
    stellar halo. arXiv:2002.12043.

6   Bromm, V. & Yoshida, N. 2011. The first galaxies. *Annual Review of
    Astronomy and Astrophysics*, vol. 49, issue 1: 373–407.

7   Greif, T. *et al.* 2008. The first galaxies: assembly, cooling and the
    onset of turbulence. *Monthly Notices of the Royal Astronomical
    Society*, vol. 387, issue 3: 1021–1036.

8   Jeon, M. *et al.* 2014. Recovery from Population III supernova
    explosions and the onset of second-generation star formation.
    Monthly Notices of the Royal Astronomical Society, vol. 444, issue 4:
    3288–3300.

9   Wise, J. & Abel, T. 2008. Resolving the formation of protogalaxies.
    III. Feedback from the first stars. *The Astrophysical Journal*, vol. 685,
    issue 1: 40–56.

10 Muratov, A. *et al.* 2013. Revisiting the first galaxies: the epoch of
    Population III stars. *The Astrophysical Journal*, vol. 773, issue 1,
    article id. 19: 9 pp.

11 Gendin, N. & Kravtsov, A. 2006. Fossils of reionization in the Local
    Group. *The Astrophysical Journal*, vol. 645, issue 2: 1054–1061.

12 Frebel, A. *et al.* 2014. Segue 1: an unevolved fossil galaxy from the
    early Universe. *The Astrophysical Journal*, vol. 786, issue 1, article id.
    74: 19 pp.

13 Vargas, L. *et al.* 2013. The distribution of alpha elements in
    ultra-faint dwarf galaxies, *The Astrophysical Journal*, vol. 767, issue
    2, article id. 134: 13 pp.

14 Webster, D. 2016. Segue 1 – a compressed star formation history
    before reionization. *The Astrophysical Journal*, vol. 818, issue 1,
    article id. 80: 11 pp.

15 Jacobson, H. & Frebel, A., 2014. Observational nuclear
    astrophysics: neutron-capture element abundances in old, metal-
    poor stars. *Journal of Physics G: Nuclear and Particle Physics*, vol. 41,
    issue 4, article id. 044001.

16 Roederer, I. 2013. Are there any stars lacking neutron-capture
    elements? Evidence from strontium and barium. *The Astronomical
    Journal*, vol. 145, issue 1, article id. 26: 6 pp.

17 Magg, M. *et al.* 2018. Predicting the locations of possible long-lived low-mass first stars: importance of satellite dwarf galaxies. *Monthly Notices of the Royal Astronomical Society*, vol. 473, issue 4: 5308–5323.

18 Scott, P. *et al.* 2010. Direct constraints on minimal supersymmetry from Fermi-LAT observations of the dwarf galaxy Segue 1. *Journal of Cosmology and Astroparticle Physics*, issue 01, id. 031.

19 MAGIC Collaboration. 2016. Limits to dark matter annihilation cross-section from a combined analysis of MAGIC and Fermi-LAT observations of dwarf satellite galaxies. *Journal of Cosmology and Astroparticle Physics*, issue 02, article id. 039.

20 Ajello, M. *et al.* 2016. Fermi-LAT observations of high-energy gamma-ray emission toward the Galactic center. *The Astrophysical Journal*, vol. 819, issue 1, article id. 44: 30 pp.

21 Spekkens, K. *et al.* 2013. A deep search for extended radio continuum emission from dwarf spheroidal galaxies: implications for particle dark matter. *The Astrophysical Journal*, vol. 773, issue 1, article id. 61: 16 pp.

22 Jeltema, T. & Profumo, S. 2016. Deep XMM observations of Draco rule out at the 99% confidence level a dark matter decay origin for the 3.5 keV line. *Monthly Notices of the Royal Astronomical Society*, vol. 458, issue 4: 3592–3596.

23 Brandt, T. 2016. Constraints on MACHO dark matter from compact stellar systems in ultra-faint dwarf galaxies. *The Astrophysical Journal Letters*, vol. 824, issue 2, article id. L31: 5 pp.

24 Bullock, J. & Boylan-Kolchin, M. 2017. Small-scale challenges to the LCDM paradigm. *Annual Review of Astronomy and Astrophysics*, vol. 55, issue 1: 343–387.

25 Homma, D. *et al.* 2019. Boötes. IV. A new Milky Way satellite discovered in the Subaru Hyper Suprime-Cam Survey and implications for the missing satellite problem. *Publications of the Astronomical Society of Japan*, vol. 71, issue 5, id.94.

26 Fattahi, A. *et al.* 2020. The missing dwarf galaxies of the Local Group. *Monthly Notices of the Royal Astronomical Society*, vol. 493, issue 2: 2596–2605.

27 Simon, J. 2019. The faintest dwarf galaxies. *Annual Review of Astronomy and Astrophysics*, vol. 57: 375–415.

Chapter 9: The Cosmic Dusk

1 www.jwst.nasa.gov.
2 www.youtube.com/watch?v=bTxLAGchWnA.
3 www.theverge.com/2018/8/1/17627560/james-webb-space-telescope-cost-estimate-nasa-northrop-grumman.
4 Surace, M. *et al.* On the detection of supermassive primordial stars – II. Blue supergiants. *Monthly Notices of the Royal Astronomical Society*, vol. 488, issue 3: 3995–4003.
5 Pawlik, A. *et al.* 2011. The first galaxies: assembly of disks and prospects for direct detection. *The Astrophysical Journal*, vol. 731, issue 1, article id. 54: 17 pp.
6 James, O. *et al.* 2015. Gravitational lensing by spinning black holes in astrophysics, and in the movie *Interstellar*. *Classical and Quantum Gravity*, vol. 32, issue 6, article id. 065001.
7 https://eventhorizontelescope.org.
8 Event Horizon Telescope Collaboration. 2019. First M87 Event Horizon Telescope Results. I. The shadow of the supermassive black hole. *The Astrophysical Journal Letters*, vol. 875, issue 1, article id. L1: 17 pp.
9 Moriya, T. *et al.* 2019. Searches for Population III pair-instability supernovae: predictions for ULTIMATE-Subaru and WFIRST. *Publications of the Astronomical Society of Japan*, vol. 71, issue 3, id.59.
10 Hartwig, T. *et al.* 2018. Detection strategies for the first supernovae with JWST. *Monthly Notices of the Royal Astronomical Society*, vol. 479, issue 2: 2202–2213.
11 Yue, B. *et al.* 2014. The brief era of direct collapse black hole formation. *Monthly Notices of the Royal Astronomical Society*, vol. 440, issue 2: 1263–1273.
12 Natarajan, P. *et al.* 2017. Unveiling the first black holes with JWST: multi-wavelength spectral predictions. *The Astrophysical Journal*, vol. 838, issue 2, article id. 117: 10 pp.
13 Bañados, E. *et al.* 2018. An 800-million-solar-mass black hole in a significantly neutral Universe at a redshift of 7.5. *Nature*, vol. 553, issue 7689: 473–476.
14 Smith, B. *et al.* 2018. The growth of black holes from Population III remnants in the Renaissance simulations. *Monthly Notices of the Royal Astronomical Society*, vol. 480, issue 3: 3762–3773.
15 Woods, T. *et al.* 2017. On the maximum mass of accreting primordial supermassive stars. *The Astrophysical Journal Letters*, vol. 842, issue 1, article id. L6: 5 pp.

16 www.ligo.caltech.edu.
17 Levin, J. 2016. *Black Hole Blues and Other Songs from Outer Space*, Vintage, London, UK.
18 Abbott, B. *et al*. 2016. Observation of gravitational waves from a binary black hole merger. *Physical Review Letters*, vol. 116, issue 6, id.061102.
19 https://lisa.nasa.gov.

Other useful references:
Bartusiak, M. 2015. *Black Hole*, Yale University Press, New Haven and London.

Chapter 10: The Epoch of Reionisation

1 Schaerer, D. 2002. On the properties of massive Population III stars and metal-free stellar populations. *Astronomy and Astrophysics*, v. 382: 28–42.
2 Ahn, K. *et al*. 2012. Detecting the rise and fall of the first stars by their impact on cosmic reionization. *The Astrophysical Journal Letters*, vol. 756, issue 1, article id. L16: 7 pp.
3 Mesinger, A. *et al*. 2013. Signatures of X-rays in the early Universe. *Monthly Notices of the Royal Astronomical Society*, vol. 431, issue 1: 621–637.
4 Fan, X. *et al*. 2001. A Survey of z>5.8 quasars in the Sloan Digital Sky Survey. I. Discovery of three new quasars and the spatial density of luminous quasars at z~6. *The Astronomical Journal*, vol. 122, issue 6: 2833–2849.
5 Giallongo, E. *et al*. 2015. Faint AGNs at z > 4 in the CANDELS GOODS-S field: looking for contributors to the reionization of the Universe. *Astronomy & Astrophysics*, vol. 578, id.A83: 14 pp.
6 Madau, P. & Haardt, F. 2015. Cosmic reionization after Planck: could quasars do it all? *The Astrophysical Journal Letters*, vol. 813, issue 1, article id. L8: 6 pp.
7 Stacy, A. & Bromm, V. 2013. Constraining the statistics of Population III binaries. *Monthly Notices of the Royal Astronomical Society*, vol. 433, issue 2: 1094–1107.
8 Xu, H. *et al*. 2014. Heating the intergalactic medium by X-rays from Population III binaries in high-redshift galaxies. *The Astrophysical Journal*, vol. 791, issue 2, article id. 110: 17 pp.

9 Robertson, B. *et al*. 2013. New constraints on cosmic reionization from the 2012 Hubble Ultra Deep Field campaign. *The Astrophysical Journal*, vol. 768, issue 1, article id. 71: 17 pp.

10 Kaurov, A. *et al*. 2016. The effects of dark matter annihilation on cosmic reionization. *The Astrophysical Journal*, vol. 833, issue 2, article id. 162: 7 pp.

11 Liu, H. *et al*. 2016. Contributions to cosmic reionization from dark matter annihilation and decay. *Physical Review D*, vol. 94, issue 6, id.063507.

12 Schön, S. *et al*. 2018. Dark matter annihilation in the circumgalactic medium at high redshifts. *Monthly Notices of the Royal Astronomical Society*, vol. 474, issue 3: 3067–3079.

13 Bromley-Davenport, J. 2013. *Space Has No Frontier: The Terrestrial life and Times of Sir Bernard Lovell*. Bene Factum Publishing, London, UK.

14 www.lofar.org.

Other useful references:

Furlanetto, S. et al. 2006. Cosmology at low frequencies: The 21 cm transition and the high-redshift Universe. *Physics Reports*, Volume 433, Issue 4–6, p. 181–301.

Loeb, A. 2010. *How Did the First Stars and Galaxies Form?* Princeton University Press, New Jersey, US.

Wise, J. 2019. An introductory review on cosmic reionization. *Contemporary Physics* 60, 2: 145–163.

Zaroubi, S. 2013. The epoch of reionization, the first galaxies. *Astrophysics and Space Science Library*, vol. 3: 45.

Chapter 11: Unknown Unknowns

1 Lorimer, D. *et al*. 2007. A bright millisecond radio burst of extragalactic origin. *Science*, vol. 318, issue 5851: 777–.

2 Petroff, E. *et al*. 2019. Fast radio bursts. *The Astronomy and Astrophysics Review*, vol. 27, issue 1, article id. 4: 75 pp.

3 Petroff, E. *et al*. 2015. Identifying the source of perytons at the Parkes radio telescope. *Monthly Notices of the Royal Astronomical Society*, vol. 451, issue 4: 3933–3940.

4 www.skatelescope.org.

5 Hoare, M. *et al.* SKA and the cradle of life. *Proceedings of Advancing Astrophysics with the Square Kilometre Array PoS(AASKA14)115.* 9 –13 June, 2014.

6 Pritchard, J. *et al.* Cosmology from EoR/Cosmic Dawn with the SKA. *Proceedings of Advancing Astrophysics with the Square Kilometre Array PoS(AASKA14)012.* 9 –13 June, 2014.

7 Burns, J. *et al.* 2019. FARSIDE: A low radio frequency interferometric array on the lunar farside, Astro2020: Decadal Survey on Astronomy and Astrophysics, APC white papers, no. 178. *Bulletin of the American Astronomical Society,* vol. 51, issue 7, id. 178.

8 Bentum, M. *et al.* 2020. A roadmap towards a space-based radio telescope for ultra-low frequency radio astronomy. *Advances in Space Research,* vol. 65, issue 2: 856–867.

9 www.isispace.nl/projects/ncle-the-netherlands-china-low-frequency-explorer.

Other useful references:

Wild, S. 2012. *Searching African Skies.* Jacana Media (Pty) Ltd., Johannesburg, South Africa.

Acknowledgements

Writing this book was hard, I won't lie. Shortly after I signed the contract I found out I was pregnant, so I ended up having not only to birth a baby but a book too ... and I'm not entirely sure which was harder. The only reason it happened at all was because of the help and love from the below.

This work would not have been possible if not for generous support from the Royal Society in the form of the Dorothy Hodgkin Fellowship, providing a flexibility to academic research that is an example to be followed. To my Bloomsbury editors, Jim Martin and Anna MacDiarmid, without whom there would be no book, thank you for taking a chance. Thank you to Léon Koopmans, Abraham Loeb and Howard Bond for being so generous with their time and answering my many questions. Thank you to Mattis Magg, Paul Clark, Joshua Simon, Katie Paterson, Manu Palomeque, Richard Porcas, NASA, ESO, ESA, Michael Goh, ICRAR and the Illustris Collaboration for their kind permission to reproduce their figures, photographs and artwork. I am forever indebted to Ryan Stimpson, Phil Parker, Brett Handley, Paul Woods and Steven Chapman for their invaluable comments and suggestions on the manu-script, thank you so much for taking the time. Jonathan Pritchard, you gave me the space to figure out what kind of a scientist to be, and I couldn't have stayed in astronomy or written this book without your support, thank you.

Steve, you have taken on far more than your fair share of cooking, cleaning and cheerleading, and without a grumble. I love you and love life because of you. Thank you. Lyra, Cassie and Olive, my three little stars. Lyra, your curiosity is inspiring and reminds me why science is so wonderful when I get too focused in on the complexities. Cassie, your passion for life reminds me to take a break and have fun. Olive, your chuckles make me explode with joy. Mum and Dad, thank you for bringing me up to see the wonder in everything and above all keep asking questions. Maxine, Sarah, Phil and Bill, thank you so much for all the babysitting, sleepovers and family dinners that you provided for the kids while I worked. Rachael, Christine, Sue, thank you for being wonderful friends and being so understanding when I complained about the intricacies of book writing, always.

And of course thank you to anyone who has bought the book and even read to the end of this part. I hope it means you enjoyed it as much as I enjoyed writing it. Because, as hard as it was, it was the most wonderful experience.

Index

Page numbers in **bold** indicate figures.

Absolute Radiometer for Cosmology, Astrophysics and Diffuse Emission 2 (ARCADE 2)   133
absorption lines
  galactic spectra   72–75, **73**
  quasar spectra   241–242, **241**
  stellar spectra   34–37, **35**, 50–52, **51**, 55–56, 61, 166–167
acceleration   89–91
accretion disks
  black holes   221–222, 224, 225, 235
  star formation   150–152, **151**
Adams, Douglas   90
airships   21, 26, 95–96
All the Dead Stars (Paterson)   126
Aller, Lawrence   168–169, 170
Alpha Centauri   43
alpha-element ratios   192–194
anaerobic life   147
Andromeda   184
  age of light from   43–44
  blueshifting   73, **73**, 74, 75
  calcium absorption lines   72–73, 74, 75
  collision with Milky Way   75, 187
  dwarf galaxies around   205
  high- and low-velocity stars   170
  H-R diagrams   60
  luminosity   186
angular size   38
Antennae Galaxies   187
Antikythera device   24
astrology   49
astronauts   90–91

asymptotic giant branch (AGB) stars   194
atomic bombs   97–98
atomic cooling halos (ACH)   190–191

Baade, Walter   60, 170
background radiation
  excess background theory   132–133
  see also cosmic microwave background
baryonic mass   189
Beanie Baby bubble   47–48
Bell Telephone Laboratory   66
Bell-Burnell, Jocelyn   53
beryllium   153
Betelgeuse   49
Big Bang   10, **11**, 18, 70, 77
  see also cosmic microwave background
binary systems   227, 236
black holes   60, 176, 218–222
  accretion disks   221–222, 224, 225, 235
  accretion limit   225
  binary systems   227
  at centre of galaxies   32, 133–134, 221
  direct collapse   223–224, 225–226, 227
  first image   220–221
  light and   219–220
  Massive Compact Halo Objects (MACHOs)   202

Population III stars   133–134,
    223–224, 235
quasars   235–236, 240–242, **241**,
    270
supermassive   134, 221, 222,
    224–226
blackbody spectrum   33–35, **34**, 78,
    79–80, 84–85, 111–112, 143
blood moons   33
blueshifting   73, **73**, 74, 75
Bond, Howard   63–64
Bowman, Judd   121
Brown, Ernest   24
brown dwarfs   105, 126–127
Bryson, Bill   205
Butler, Howard   26–27

calcium   72–73, 74, 75
Cannon, Annie Jump   50–51
carbon   98, 153, 218
carbon dioxide   140, 141
Carter, Howard   160, 170–171
Cassini, Giovanni Domenico   41
Centaurus   32
Chamberlain, Joseph   168–169, 170
Chang'e missions   273–274
chess   80–81
Columbus, Christopher   24
core-collapse supernovae   176, 193
corona   22, 26–28, 36–37, 39–40
coronagraph   39
Coronal Heating Problem   27–28,
    37–38
coronium   22, 36–37
Cosmic Background Explorer
    (COBE)   79, 239
Cosmic Dawn   110, 215–217, 227,
    228, 230, **230**, 264, 275
    see also Experiment to Detect the
    Global EoR Signature
    (EDGES)
cosmic microwave background
    84–85, 113, 231
    constraint on Epoch of
    Reionisation   239–240
    discovery   66–70, 78–80

spin temperature and   114, 115,
    117–119, **117**
Cosmic Spectrum, The (Pater-
    son)   109, 112
cosmological constant   70
cyanobacteria   139–140, 141, 147,
    148, 156–157

Dark Ages   11, **11**, 110, 116, 119,
    130, 135–136, 230, **230**,
    263–264, 271–272, 274–275
Dark Energy Survey   **197**, 204
dark matter   121–131, 183, 189, 275
    annihilation   200–201, 237–238
    in dwarf galaxies   199–202
    EDGES results and   128–131
    flat rotation curves and   124–125,
    126
    halos   123, 124
    Massive Compact Halo Objects
    (MACHOs)   202
    millicharged dark matter
    model   130–131
    minihalos   190–191, 234, 238
    reionisation of hydrogen
    and   237–238
    Weakly Interacting Massive
    Particles (WIMPs)   127–128,
    200–201
de Bruyn, Ger   256
deuterium   83, 97, 98, 99
Dicke, Robert   78–79
direct collapse black holes
    (DCBHs)   223–224, 225–226,
    227
Doppler effect   71–76, **73**, 124
Durham, UK   204–205, 207
dwarf galaxies   181–209
    alpha-element ratios   192–194
    dark matter   199–202
    defining   188–189
    differentiating from globular
    clusters   182, 183, 188, 189
    dwarf spheroidals   185–186
    Eridanus II   202
    fossil first galaxies   191–196

galactic cannibalism   183–184,
     186–188
Large Magellanic Cloud   184,
     185
mass   183, 188–189
metallicity   189
metal-poor stars   192–196
missing satellite problem
     203–204
naming   182
neutron-capture elements
     194–195, 196
numbers   **197**, 203–205
probability of finding Population
     III stars in   196–199
reionisation of hydrogen
     and   237–238
Segue   1 181, 182, 183–184,
     192–196, 199, 200–201, 203,
     206
Small Magellanic Cloud   184,
     185
stellar chemistry   192–196
ultra-faint   185–186, 188,
     199–202
dynamical mass   189

Earth
     Great Oxygenation Event
          140–141, 147, 148
     orbit   124
     snowball Earth   147–148
     temperature   140
Earth-Moon-Earth (Paterson)   136
eclipses
     lunar   23–24, 33
     solar   21–28, 36, 38–39, 53–54
Eddington, Arthur   53–54
Edison Company   21, 24
Egypt, ancient   49, 160–161,
     163–164, 170–171
Einstein, Albert   31, 57, 70, 98–99,
     226
elastic energy   92
electromagnetic spectrum   **23**, 29
     see also light

electrons   78, 83–84, 145, 201
     ionisation   55, **230**, 231, 232–233
     pair production   222
     spin-flip transitions   113–115,
          113
Emerson, Ralph Waldo   9
endowment effect   48
energy
     conservation   92–93
     elastic   92
     force and   91–92
     gravitational potential   92–94, 144
     kinetic   89, 92, 93, 94, 145
     mass and   98–99
     thermal   92, 93, 144–145
Epoch of Reionisation   115,
     229–243, 258–259, 264,
     269–270, 275
     constraints   239–243
     cosmic microwave background
          and   239–240
     dwarf galaxies   237–238
     LOFAR-Epoch of Reionisation
          project   255–258
     Population III stars   233–235,
          233, 238, 243
     quasars   235–236, 240–242, **241**,
          270
     reionisation of hydrogen   **230**,
          231, 232–238, **233**, 239–243, **241**
     see also Experiment to Detect the
          Global EoR Signature
          (EDGES)
Eridanus II   202
Event Horizon Telescope
     (EHT)   220–221
excess background theory   132–133
Experiment to Detect the Global
     EoR Signature
     (EDGES)   115–121, **117**, 137,
          257–258, 264
     absorption trough   118–121, **120**,
          134–135
     dark matter theories   128–131
     early presence of black
          holes   133–134

excess background theory
 132–133
foregrounds theory 134–135
exponential growth 80–81

Farside Array for Radio Science
 Investigations of the Dark Ages
 and Exoplanets (FAR-
 SIDE) 273
Fast Radio Bursts (FRBs) 261
Fermi Large Area Telescope
 (Fermi-LAT) 200–201
first galaxies 122–123, 189–196
first stars see Population III stars
First Three Minutes, The (Wein-
 berg) 83
flat rotation curves 124–125, **126**
Fleming, Williamina P. 50
Florian Goebel Major Atmospheric
 Gamma-ray Imaging Cheren-
 kov (MAGIC) 200–201
Frost, Edwin 49
Fukushima disaster, Japan 101–102

galaxies
 atomic cooling halos
  (ACH) 190–191
 black holes at centre 32,
  133–134, 221
 Centaurus A 32
 collisions 75, 187–188
 definitions 181–182
 Doppler effect 71–76, **73**, 124
 first 122–123, 189–196
 galactic cannibalism 183–184,
  186–188
 Hubble flow 76
 Local Group 184–185
 M87 221
 mass-to-light ratio 183
 spectra 72–75, **73**
 speed 76
 spiral 60
 see also Andromeda; dwarf galaxies;
  Milky Way
Galilei, Galileo 41
gamma rays 200–201

Gamow, George 45
gas clouds see star formation
gas pressure 100, **101**, 145
general relativity 70, 226, 266
Giant Metrewave Radio Telescope
 (GMRT) 252
Glacier National Park, US 157
globular clusters 172, 182, 183,
 188, 189
gravitational potential energy
 92–94, 144
gravitational waves 226–228
gravity 88–95, 100, **101**, 106–107
Great American Eclipse 27
Great Oxygenation Event
 140–141, 147, 148
Greece, ancient 49
Green Bank Telescope 248–249

Halley, Edmond 24
Hamelin Pool, Australia 156–157
height, human 162–164
helium 14, 36, 83–84, 96, 98, 99,
 112, 153, 192–193
Herschel, Caroline 30
Herschel, William 30
Hertzsprung-Russell (H-R)
 diagrams 58–61, **59**, 170
Hill-Brown theory 24
Hindenburg airship 95–96
Hiroshima, Japan 97
Hitchhiker's Guide to the Galaxy, The
 (Adams) 90
Holmdel Horn Antenna 66–70,
 78–79, 84
Holst, Gustav 54
Hubble, Edwin 76
Hubble flow of galaxies 76
Hubble Space Telescope 10, 169,
 214, 215, 216, 237, 243–244
Hubble Ultra Deep Field 215, 237
Hubble–Lemaître law 76
hydrogen 14, 56, 57, 61, 95, 112, 275
 21cm radiation 113–119, **117**,
  251–252, **251**
 absorption lines 50, 51, 241–242,
  241

in airships 95–96
collisions 145–146
flammability 95–96
isotopes 83, 97–98
molecular hydrogen
  cooling 146–147, **146**,
  148–149, 190–191, 223
nuclear fusion 97–100, **100**,
  102–103, 104–105, 106–107,
  153
nucleosynthesis 83–84
recombination 84, 234, 238
reionisation **230**, 231, 232–238,
  233, 239–243, **241**
spin temperature 114–115,
  117–119, **117, 120**
see also star formation
hydrogen bombs 98

Iben, Icko, Jr. 176, 177
ice age 147–148
Illustris simulation 123
inflation period 80–81
infrared radiation 29, 30, 31, 216
initial mass function 161–165, 178,
  195, 214–215
interferometry 249–250
International Dark Sky
  Places 111
International Thermonuclear
  Experimental Reactor
  (ITER) 105
Interstellar (film) 219
interstellar medium 112
  metal enrichment 154–155, 177,
  178, 190–191, 193, 195
Io 41–43, **42**
ionisation 55–56
  reionisation of hydrogen **230**,
  231, 232–238, **233**, 239–243,
  241
ionosphere 66, 246, 247, 272
iron 36–37, 153, 166–167,
  193, 218
  see also metal-poor stars
ironstones 141
isotopes 83, 97–98

James Webb Space Telescope (JWST)
  211–217, 224, 226, 263
Japan
  Fukushima disaster 101–102
  Nagasaki and Hiroshima
    bombings 97
  Naha tug-of-war 87–88, 91–92
Jeans, Sir James Hopwood 143
Jeans mass 143–144, 147, 150, 151,
  152, 223
Jupiter, moons of 41–43, **42**

Keplerian descent 124–125
kinetic energy 89, 92, 93, 94, 145
King, Ivan 64
Koopmans, Leon 255–258

Lagrange points 213–214, **213**
Lake District, UK 139
Lake Erie, US 139–140
Large Magellanic Cloud 184, 185
Large Synoptic Survey Telescope
  (LSST) **197**, 204
Laser Interferometer Gravitational-
  Wave Observatory
  (LIGO) 226–227
Laser Interferometer Space Antenna
  (LISA) 227–228, 263
Lemaître, Georges 76
light
  black holes and 219–220
  colours 29–30
  Doppler effect 71–76, **73**, 124
  electromagnetic spectrum **23**, 29
  infrared 29, 30, 31, 216
  mass-to-light ratio 183
  speed 40–44, **42**, 99
  ultraviolet (UV) 31, 115,
    118–119, 148, 233–235, **233**,
    237–238
  wavelengths 28–32
  wave-particle duality 30–31
  see also photons
light pollution 111
lightning 44
lithium 83
Local Group of galaxies 184–185

Lockyer, Norman   36
Loeb, Abraham   127
Long Wavelength Array (LWA)   133
longitude   40–41
Lovell, Bernard   246–247
Lovell Telescope   247–248
Low Frequency Array
    (LOFAR)   249–250, 251,
    255–258, 266, 267, 272
luminous mass   124, 182–183, 189
lunar eclipses   23–24, 33
Lunar Low Frequency Antennas for
    Radio Astronomy
    (LUFAR)   273
Lyman-Werner band of ener-
    gies   148

M87   221
magnesium   193
Magpie Bridge satellite   274
main sequence stars   59, **59**
mass
  baryonic   189
  deaths of stars and   59–60, **59**,
    217–226
  dynamical   189
  energy and   98–99
  gravity and   88–89
  Jeans mass   143–144, 147, 150,
    151, 152, 223
  luminous   124, 182–183, 189
  mass-lifetime relation   162, 164
  mass-to-light ratio   183
  maximum mass of stars   105–106
  measuring   182–183, 188–189
  star formation and   105–106,
    143–144, 147, 150–152
Massive Compact Halo Objects
    (MACHOs)   202
Mercury   124
metal content of stars   14–16, 18,
    36–37, 52, 56, 61, 99, 149,
    152–155
  in dwarf galaxies   186, 189, 193
  metal-poor stars   166–168, **167**
metal cooling   147, 149

metal enrichment of interstellar
    medium   154–155, 177, 178,
    190–191, 193, 195
metal-free stars see Population III
    stars
metal-poor stars   62, 166–179
  discovery   61, 168–170
  in dwarf galaxies   192–196
  metallicity values   166–168, **167**
  in Milky Way   63, 170–177, 196
  searching for   63, 170–177,
    192–196
  SM0313-6708   175–176
  SMSS J1605-1443   174, 175, 176
  velocities   170, 173
methane   140, 141, 147
Mice Galaxies   187
microwave ovens   262
microwaves   29
Milky Way   184–185
  black hole at centre   133–134,
    221
  bulge and disk   172
  collision with Andromeda   75,
    187
  gamma rays   201
  halo   67, 170, 172–173, 196
  high- and low-velocity stars   170
  H-R diagrams   60
  metal-poor stars   63, 170–177,
    196
  missing satellite problem
    203–204
  numbers of dwarf galaxies
    around   203–205
  probability of finding Population
    III stars in   198
  Sagittarius Stream   188
millicharged dark matter
    model   130–131
minihalos   190–191, 234, 238
missing satellite problem   203–204
Mitchell, Edgar   91
molecular hydrogen cooling
    146–147, **146**, 148–149,
    190–191, 223

Moon
    acceleration experiment
        on   90–91
    angular size   38
    Chang'e missions   273–274
    distance from Earth   39
    lunar eclipses   23–24, 33
    radio arrays on   272–274
    solar eclipses   21–28, 36, 38–39,
        53–54
    tidal locking   273
moons of Jupiter   41–43, **42**
mummies   160–161, 164, 171,
    174–175
Murchison Widefield Array
    (MWA)   252

Nagasaki, Japan   97
Naha tug-of-war, Japan   87–88,
    91–92
nanoflares   37, 40
National Aeronautics and Space
    Administration (NASA)   26,
    27, 39–40, 90–91, 170
National Air and Space Museum,
    Washington DC   69
neon   153, 193
Neptune   124, 125
Netherlands-China Low Frequency
    Explorer (NCLE)   256, 274
neutron stars   60, 194, 217–218
neutron-capture elements   1
    94–195, 196
neutrons   78, 82, 83–84, 98, 101
New York Times   24
Newton, Isaac   29–30, 31
Nobel Prize for Physics   79
Notes From a Small Island
    (Bryson)   205
nuclear fission   97–98, 101–102
nuclear fusion   97–100, **100**,
    102–103, 104–105, 106–107,
    153
nuclear power   101–103, 104–105
nuclear weapons   97–98
nucleosynthesis   81–84

Olbers' paradox   9
Oort, Jan   170
optical astronomy   244, 248
Orion   49
oxygen   96, 153, 192–193, 218, 223
    see also Great Oxygenation Event

pair production   222
pair-instability supernovae   193,
    222–223
Parker Solar Probe   39–40
Parkes radio telescope   261
Paterson, Katie   109, 111, 112, 122,
    126, 136
Payne-Gaposchkin, Cecilia   52–57,
    58, 61
Peebles, P. James   78–79
Penzias, Arno Allan   66–70, 78–79,
    84, 113, 239
perytons   261–262
Peterson, Alvin   21, 22, 26, 36
Petrarch   135
photons   30–31, 78, 84–85, 94, 145
    molecular hydrogen
        cooling   146, **146**, 148–149,
        190–191
    pair production   222
    reionisation of hydrogen   **230**,
        231, 232–238, **233**, 239–243,
        241
    spin-flip transitions   113–115, **113**
    see also light
photosynthesis   141
Pickering, Edward C.   50
pigeons   65–66, 68–69
Pink Floyd   29
Planck telescope   80, 113, 213, 239,
    240
planetary nebulae   217
plasma   104–105
plutonium   98
Poe, Edgar Allan   9–10
Polynesia   49
Population I stars   15, 60–61, 62, 72
Population II stars   15, 60–61, 62,
    72, 154, 191, 237

Population III stars   11–16, **11**, 18,
    62–64, 72, 141–142, 147
  black holes   133–134, 223–224,
    235
  brief lives   152–154
  dark matter and   122–123
  deaths   222–224, 228
  first detection   115–121, **120**
  first galaxies   122–123, 189–196
  formation   142, 145–147,
    150–152, **151**, 156
  initial mass function   164–165,
    178, 214–215
  mass   106, 147
  probability of finding   196–199
  reionisation of hydrogen
    233–235, **233**, 238, 243
  supernovae   133, 153–155, 190,
    191, 222–223, 234
  surface temperature   148
  X-ray binaries   236
  see also radio astronomy; stellar
    archaeology
positrons   201, 222
Precision Array for Probing the
  Epoch of Reionization
  (PAPER)   252
prisms   29–30
proton–proton chain reaction   99,
  100
protons   78, 82, 83–84, **230**, 231
protostars   94, 105, 150–152, **151**,
  223

quantum mechanics   30–31, 53
quasars   235–236, 240–242, **241**, 270
Queqiao satellite   274

radar   245–247
radio astronomy   243–259, 261–275
  Fast Radio Bursts (FRBs)   261
  foreground mitigation   253–254
  interferometry   249–250
  ionosphere and   272
  LOFAR-Epoch of Reionisation
    project   255–258

Low Frequency Array
  (LOFAR)   249–250, 251,
  255–258, 266, 267, 272
noise foregrounds   252–253
perytons   261–262
radio arrays on Moon   272–274
search for   21cm radiation
  251–252
Square Kilometre Array   256,
  265–271, **268**, 272, 275
radio waves   29, 32, 66–70, 78–80
recombination   84, 234, 238
red giants   59, **59**, 60, 110, 192, 198,
  199
redshifting   **73**, 74, 75–76, 110, 118,
  251–252, **251**
reindeer   31
relativity   53–54, 70, 159, 226, 266
Roll, Peter   78–79
Roman, Nancy   169–170, 173
Rømer, Ole   40, 41–43
Rowe, Albert Percival   246
Rubin, Vera   124–125, **126**
Russell, Henry   56, 57, 60–61
Rutherford, Ernest   48, 53, 192
Ryle, Martin   77

Sagan, Carl   61
Sagittarius Stream   188
San people   49
Secchi, Angelo   50
Segue   1 181, 182, 183–184,
  192–196, 199, 200–201, 203,
  206
Shark Bay, Australia   156–157
Shepard, Alan   91
silicon   153, 193, 218, 223
Sloan Digital Sky Survey   **197**,
  204
'Small Galaxies, Cosmic Questions'
  conference   207–208
Small Magellanic Cloud   184, 185
snakes   31
snowball Earth   147–148
solar eclipses   21–28, 36, 38–39,
  53–54

solar flares   37, 40
solar spectrum   32–37, **34, 35**
Solar System   24–25, 124, 125
   see also Sun
solar wind   39
special relativity   159
spectra
   blackbody   33–35, **34**, 78, 79–80,
      84–85, 111–112, 143
   galaxies   72–75, **73**
   quasars   240–242, **241**
   solar   32–37, **34, 35**
   stellar   32–37, **34, 35**, 50–52,
      51, 55–56, 61, 166–167,
      168–169
spin temperature   114–115,
   117–119, **117, 120**
spin-flip transitions   113–115, **113**
Square Kilometre Array   256,
   265–271, **268**, 272, 275
Star Called The Sun, A (Gamow)   45
star formation   87–107, 112
   accretion disks   150–152, **151**
   combustion   95–97
   gas pressure   100, **101**, 145
   gravitational contraction   88–95,
      100, **101**, 107, 143–145
   Jeans mass   143–144, 147, 150,
      151, 152, 223
   mass and   105–106, 143–144,
      147, 150–152
   metal cooling   147, 149
   molecular hydrogen
      cooling   146–147, **146**,
      148–149, 190–191, 223
   nuclear fusion   97–100, **100**,
      106–107
   protostars   94, 105, 150–152, **151**,
      223
   stellar nurseries   142–143,
      149–152, **151**, 162
   temperature and   143–147,
      148–149, 150
stars
   absorption lines   34–37, **35**,
      50–52, **51**, 55–56

asymptotic giant branch (AGB)
   stars   194
brown dwarfs   105, 126–127
classification   49–52, **51**,
   55–56
colour and luminosity
   57–59, **59**
deaths   59–60, **59**, 217–226
H-R diagrams   58–61, **59**, 170
main sequence   59, **59**
mass-lifetime relation   162, 164
maximum mass   105–106
neutron stars   60, 194, 217–218
plasma   104
Population I stars   15, 60–61, 62,
   72
Population II stars   15, 60–61, 62,
   72, 154, 191, 237
red giants   59, **59**, 60, 110, 192,
   198, 199
stellar spectra   32–37, **34, 35**,
   50–52, **51**, 55–56, 61
temperature   32–34, 37
velocities   124–125, 170, 173
white dwarfs   59, **59**, 217
   see also metal-poor stars; star
      formation; stellar archaeology;
      Sun; supernovae
steady state theory of Universe   77,
   143
stellar archaeology   159–161,
   177–179
   alpha-element ratios   192–194
   discovery of metal-poor stars   61,
      168–170
   initial mass function   161–165,
      178, 195, 214–215
   neutron-capture elements
      194–195, 196
   probability of finding Population
      III stars   196–199
   searching for metal-poor
      stars   63, 170–177, 192–196
stellar spectroscopy   50, 192
   see also spectra
stromatolites   156–157

Sun
    absorption lines   34–37, **35**
    age   94
    age of light from   43, 45
    angular size   38
    classification   51
    composition   56, 61
    corona   22, 26–28, 36–37, 39–40
    density   103, 104
    on main sequence   59
    in myth   49
    nanoflares   37, 40
    nuclear fusion   99–100
    plasma   104
    solar eclipses   21–28, 36, 38–39,
        53–54
    solar spectrum   32–37, **34, 35**
    solar wind   39
    temperature   27–28, 34, 37, 100,
        103, 104
supermassive black holes
        (SMBHs)   134, 221, 222,
        224–226
supernovae   60, 167, 172, 193, 195
    core-collapse   176, 193
    dark matter and   131
    neutron stars and   218
    pair-instability   193, 222–223
    Population III stars   133,
        153–155, 190, 191, 222–223,
        234
    Type   1a   193
synchrotron radiation   133,
        252–253, 254

thermal energy   92, 93, 144–145
thermal equilibrium   78, 82
thermonuclear devices   98
Thomson optical depth   240
thunder   44
tidal locking   273
Triangulum Galaxy   184
tritium   97–98, 99
tug-of-war   87–88, 91–92
Tutankhamun   160–161, 164, 171

Ultra Deep Field   215, 237
ultra-faint dwarf galaxies   185–186,
        188
    dark matter   199–202
    Segue   1 181, 182, 183–184,
        192–196, 199, 200–201, 203,
        206
ultraviolet (UV) light   31, 115,
        118–119, 148, 233–235, **233**,
        237–238
Universe   10–16, **11**, 18, 65–86
    Big Bang   10, **11**, 18, 70, 77
    colour   109–110, 111–112
    expansion   70, 75–77, 80–81, 110
    inflation   80–81
    measuring temperature   111–121,
        117, 120
    nucleosynthesis   81–84
    recombination   84
    steady state theory   77, 143
    see also cosmic microwave
        background
uranium   98
USS Los Angeles airship   21, 26,
        96

vitamin D deficiency   17

wave-particle duality   30–31
Weakly Interacting Massive
        Particles (WIMPs)   127–128,
        200–201
Weinberg, Steven   83
white dwarfs   59, **59**, 217
Wilkinson, David   78–79
Wilkinson Microwave Anisotropy
        Probe (WMAP)   79, 80, 113,
        213, 239
Wilson, Robert Woodrow   66–70,
        78–79, 84, 113, 239
Woolf, Virginia   27
Wouthuysen-Field effect   115, 119

X-ray binaries   236
X-rays   29, 32, 133, 235–236